T0135628

Functional calculus for bisectorial operators and applications to linear and non-linear evolution equations

DISSERTATION
zur Erlangung des Doktorgrades Dr. rer. nat.
der Fakultät für Mathematik und Wirtschaftswissenschaften der
Universität Ulm

vorgelegt von MARKUS MICHAEL DUELLI aus LAUPHEIM

Januar 2005

Bibliografische Information Der Deutschen Bibliothek

Die Deutsche Bibliothek verzeichnet diese Publikation in der Deutschen
Nationalbibliografie; detaillierte bibliografische Daten sind im Internet über
http://dnb.ddb.de abrufbar.

ISBN 3-8325-0862-7

Logos Verlag Berlin
Comeniushof, Gubener Str. 47,
10243 Berlin
Tel.: +49 030 42 85 10 90
Fax: +49 030 42 85 10 92
INTERNET: http://www.logos-verlag.de

Gutachter: 1. Prof. Dr. Wolfgang Arendt
             2. Prof. Dr. Lutz Weis (Karlsruhe)
             3. PD Dr. Ralph Chill

Dekan: Prof. Dr. Ulrich Stadtmüller

Tag der mündlichen Prüfung: 16.02.2005

Contents

Introduction

The theory of differential equations in Banach spaces plays an important rôle in analysis. In fact, many partial differential equations may be rewritten as an ordinary differential equation taking values in an infinite dimensional space. A prominent example is, as in the finite dimensional case, the linear first order equation also referred to as the abstract Cauchy problem. We will describe this equation in the following.

Given an interval $I \subset \mathbb{R}$, a Banach space X, a closed and densely defined linear operator A with domain $\mathcal{D}(A)$ and an inhomogeneity $f \in L_p(I, X)$ one is interested in the existence of solutions of the equation

$$u' + Au = f. \tag{1}$$

The choice $A = -\Delta$, i.e. $-A$ is a realization of the Laplacian, gives the prototype example, the heat equation.

Let $I = [0, 1]$. We say that A has *maximal L_p-regularity*, if there is a unique solution u of (1) satisfying the initial condition $u(0) = 0$ such that u' and Au have the same regularity as the inhomogeneity f, i.e. $u \in W^{1,p}(I, X) \cap L_p(I, \mathcal{D}(A))$. Similarly, one may ask for existence of solutions on the whole real line, i.e. $I = \mathbb{R}$. In this case there is no initial condition at 0; one may regard the decay of the solution at infinity as a boundary condition.

Instead of looking for solutions in L_p spaces one may also consider spaces of Hölder continuous functions. This leads to the notion of maximal C^α-regularity. In this thesis we restrict ourselves to the case of Lebesgue spaces. Therefore, when we speak of maximal regularity, it is always meant in the L_p space sense.

The interest in this property stems from the fact that it implies the following a priori estimate

$$\|u'\|_{L_p} + \|Au\|_{L_p} \le C \|f\|_{L_p} .$$

The existence of such an estimate renders iteration arguments applicable. Hence, it is possible to obtain solutions to quasilinear differential equations of the form

$$u'(t) + A(u(t))u(t) + F(t, u(t)) = f(t), \tag{2}$$

by reformulating the equation as a *fixed point problem*; here $A(\cdot)$ is a family of autonomous operators with common domain satisfying a local Lipschitz continuity condition. For example, we will consider differential operators of the form $A(u) = a(\cdot, u)\Delta u$ for suitable

1

multiplicative perturbations a.

Studying maximal regularity on the interval $[0, 1]$ or the half-line $[0, \infty)$ leads to
sectorial operators. In contrast to this fact, maximal regularity on the line gives rise to
bisectorial operators. An invertible operator A is called bisectorial, if both iA and $-iA$
are sectorial operators, i.e. if the imaginary axis is contained in its resolvent set and if
the set $\{sR(is, A) : s \in \mathbb{R}\}$ is bounded.

Remarkably, it is (essentially) only true in Hilbert spaces that every (bi)sectorial
operator satisfies maximal regularity, as has been shown recently by Kalton and Lancien
[73]. Therefore, having realized the importance of maximal regularity for applications,
one is led to search for suitable criteria ensuring this property for a single operator. To
this end there are two elegant approaches available. One approach consists of regarding
the solution operator as a Fourier multiplier. The proof of its boundedness relies on the
recent extension of the *Mikhlin multiplier theorem* to vector valued functions due to Weis
[121, 122]. This yields that R-bisectorial operators satisfy maximal regularity on the line.

The other approach is the *operator-sum method*. Indeed, the main idea of this approach lies in the important observation that, writing $B = \frac{d}{dt}$ for the derivation operator
on $L_p(\mathbb{R}, X)$ with domain $W^{1,p}(\mathbb{R}, X)$ and identifying A with the operator induced on
$L_p(\mathbb{R}, \mathcal{D}(A))$ by pointwise multiplication, maximal regularity is equivalent to the invertibility of the operator sum $A + B$ with domain $\mathcal{D}(A + B) = \mathcal{D}(A) \cap \mathcal{D}(B)$. In order to
establish the invertibility of this sum, it is desirable to find an estimate

$$\|Ax\| + \|Bx\| \leq C \|Ax + Bx\|, \qquad x \in \mathcal{D}(A + B). \tag{3}$$

The operator-sum method goes back to the seminal work by Da Prato and Grisvard
[39] where they proved that we have maximal regularity for sectorial operators in real
interpolation spaces and, moreover, that, given two commuting sectorial operators (such
that the sum of their types is less than π), their sum $A + B$ is a closable operator. It was
a great step forward, when Dore and Venni [48] showed that, if A and B admit bounded
imaginary powers (such that the sum of the BIP-types is less than π), if A is invertible
and if the Banach space satisfies the geometrical property UMD, then the sum $A + B$
is closed and invertible. If X is a UMD space, then the derivative admits a bounded
functional calculus and in particular bounded imaginary powers of type $\pi/2$.

In the last years a powerful and versatile tool was developed that allows to derive
the inequality (3): the *holomorphic functional calculus*. Without going yet into technical
details, the functional calculus allows to give meaning to expressions like $f(B)$ where
B is a sectorial operator and the function f is holomorphic in a neighbourhood of the
spectrum. If the operator $f(B)$ is bounded for all bounded holomorphic functions, the
functional calculus is called *bounded*. To deduce the inequality (3) we would like to insert
the operator valued function $F(z) = z(z + A)^{-1}$ into the functional calculus for B; we
assume that A and B commute. We would expect (or at least hope) that the obtained
operator $F(B)$ should equal $B(A + B)^{-1}$ (if $A + B$ is invertible). Under quite natural
assumptions this is indeed the case. It is due to Kalton and Weis that the operator

$F(B)$ is bounded, if B has a bounded (scalar valued) functional calculus and if the holomorphic function F has R-bounded range. For our special function F this reduces to the R-sectoriality of $-A$.

If we want to apply the operator sum method to establish maximal regularity on the line, we are thus forced to establish a functional calculus for bisectorial operators and an extension to operator valued functions. It is in this context, that the notion of an R-bisectorial operator will appear.

The construction of a holomorphic calculus for sectorial operators goes back to McIntosh [90] in the setting of Hilbert space, and [37] in Banach space. The sectoriality assumption appears naturally, since in a first step the operator $f(A)$ is defined for functions of regular decay by means of the Cauchy-type integral

$$f(A) = \frac{1}{2\pi i} \int_\Gamma f(z) R(z, A) \, dz,$$

where Γ is the boundary of a suitable sector. The integral is forced to converge by means of a suitable growth condition on f in correspondence with the growth of the resolvent. For recent accounts on the 'state of the art' see [42], [81] and [61] and the references therein. The construction of the functional calculus can also be carried out in the setting for bisectorial operators [3], [50], [53].

Not every (bi)sectorial operator has a bounded functional calculus, not even in Hilbert space [91]. However, by now it has been established that large classes of elliptic differential operators with rather general coefficients and boundary values, Schrödinger operators with singular potentials and many Stokes operators have a bounded H^∞-calculus (see [81] and the references therein). This large pool of examples makes a stark contrast to earlier attempts to develop a spectral theory on Banach spaces, more closely modelled after the theory for normal operators on Hilbert spaces.

Second order quasilinear equations on the line ask, of course, for a second order maximal regularity result. Again, to this end one may apply either the Mikhlin-Weis theorem or the operator-sum method. Alternatively, one may rewrite the second order equation as a first order system. The operator matrix obtained in this way turns out to be a bisectorial operator.

In comparison with sectorial operators, bisectorial operators are much less studied in the literature. Therefore, it is interesting to know if one can *decompose* the underlying Banach space X in such a way, that the operator A can be written as a direct sum of two sectorial operators; this allows to transfer many results that hold true for sectorial operators to the setting of bisectorial operators.

In this thesis we will proceed along the lines sketched above. It consists of three chapters. In the first chapter we introduce the notion of a multisectorial operator. It generalizes both sectorial operators and the bisectorial operators introduced in the preceding paragraph. Then we will construct a joint operator-valued functional calculus for a finite number of commuting multisectorial operators. The construction of the holomorphic calculus is well known for sectorial and partially for bisectorial operators. We will

4 Introduction

establish analogous results in the new more complicated setting. This is the content of
Sections 1.2 – 1.5.

In Section 1.7 we establish the fact that, if A has a bounded scalar functional calculus,
it has also a bounded operator valued functional calculus for holomorphic functions with
R-bounded range. The result is due to Kalton and Weis [74] for sectorial operators and
can be found in [50] for bisectorial operators. It is then applied to derive a closed-sum
theorem. In Subsection 1.7.2 we introduce the notion of an *asymptotically bisectorial*
operator, which generalizes the notion of a bisectorial operator to more complicated
spectra. We also establish a version of the Kalton-Weis theorem and deduce a closed-
sum theorem in this setting.

In Sections 1.8 and 1.9 we recall the notion of *bounded imaginary powers*. We will
combine it with a version of Dore's theorem assuring the boundedness of the functional
calculus in real interpolation spaces in order to find another proof of the fact that, in
Hilbert space, the boundedness of the functional calculus is equivalent to the boundedness
of imaginary powers. Moreover, based on an example of Kalton, we will construct an
example of a bisectorial operator A that has a bounded sectorial functional calculus, but
whose bisectorial functional calculus is unbounded.

In [75] Kalton and Weis generalized the notion of a square function introduced by
McIntosh in the Hilbert space setting to general Banach spaces. Combined with the
notion of an *almost sectorial operator* introduced by Kalton, Kunstmann and Weis [72]
square functions provide an elegant characterization of the boundedness of the functional calculus. In Section 1.10 we extend these results to the setting of multisectorial
operators. The results are quite analogue; if we know that A has a bounded sectorial
$H^\infty(\Sigma)$-functional calculus and if the set $\{zAR(z,A)^2 : z \in \Sigma'\}$ is R-bounded, where Σ'
is a multisector contained in Σ, then A has a bounded multisectorial $H^\infty(\Sigma')$-functional
calculus. This gives at once new insight into several phenomena, e.g. we find that a
bisectorial operator A has a bounded bisectorial functional calculus if and only if iA and
$-iA$ admit bounded sectorial functional calculi – without any condition on the geometry
of the Banach space (see Corollary 1.10.17).

In the second chapter we study the spectral decomposition of multisectorial operators.
By an iteration argument it suffices to consider bisectorial operators. Given a bisectorial
operator A, the associated spectral projection $p(A)$ may turn out to be unbounded; even
in Hilbert space. In fact, we construct a counter-example on each Banach space that
admits a *Schauder basis*. In Section 2.2 we characterize the existence of the spectral decomposition by the boundedness of the projection $p(A)$, where p is the indicator function
of a suitable sector. If $0 \in \rho(A)$, it is already shown by Dore and Venni [49] that one may
characterize the boundedness of the spectral projection in terms of fractional powers of
A. Using the machinery of the functional calculus developed in the first chapter, we find
more direct and transparent proofs of these results that avoid the assumption $0 \in \rho(A)$.
Using these results we construct an example of a bisectorial operator A that has bounded
imaginary powers but BIP-type greater or equal to π and further interesting properties.
An example of such an operator has already been found by Haase [62] by different means.

In Section 2.3 we derive and generalize some classical inequalities in analysis by the boundedness of the spectral projection and the joint functional calculus. For instance, the generator of the translation group $A = \frac{d}{dt}$ on $L_p(\mathbb{R}, X)$ is a bisectorial operator. The boundedness of the associated spectral projection is equivalent to the boundedness of the Hilbert transform $(-\Delta)^{1/2}(\frac{d}{dt})^{-1}$.

In Section 2.4 this is generalized to generators A of bounded C_0-groups. In a first part, assuming the invertibility of A, we show how to deduce this from results established by Monniaux [97] (this approach is based on the *analytic generator* of a group). Then, we give a direct argument based on the functional calculus that does not require the invertibility of the generator A.

In Sections 2.5, 2.6 we consider the spectral decomposition of multisectorial operators and illustrate how to make use of the boundedness of the spectral projection in order to transfer well known results from the sectorial setting to the setting of multisectorial operators.

In the third chapter we study maximal regularity for the first and second order Cauchy problem and its applications to nonlinear equations. The first two sections are devoted to the study of maximal regularity on the line for the first and second order Cauchy problem. For the first order problem this is well known [121, 122]. The second order result was obtained by Schweiker [108] by means of the Mikhlin-Weis theorem. We give two alternative proofs of this result; one is based on the operator-sum method and a second approach via reduction of the second order equation to a first order system. The reduction method allows to deduce more precise regularity properties of the solution without making explicit use of interpolation techniques. Furthermore, we apply the closed-sum theorem in the setting of asymptotically bisectorial operators recapturing in this way a maximal regularity result for the *periodic Cauchy problem* established by Arendt and Bu [9] by means of discrete multiplier theorems.

In the third section we apply these maximal regularity results in order to prove abstract theorems establishing the existence and uniqueness of solutions of quasilinear equations (2) on the line by means of a fixed point argument. Similar results for equations on a finite interval were already obtained in [27, 30, 103]. We also give several examples that illustrate how these results apply to various concrete differential equations. For example, we show existence and uniqueness of solutions of the parabolic quasilinear equation $u_t \pm a(\|\nabla u\|_{L_2})\Delta u = g$ or the elliptic quasilinear equation $u_{tt} + a(x, u(t, x))\Delta u = g$ on cylindrical domains $\mathbb{R} \times \Omega$.

Another application of maximal regularity is given in Section 3.6 where we make use of it in order to transfer an existence theorem on center manifolds due to Mielke [93] from Hilbert spaces to the setting of UMD spaces.

Acknowledgements

At this point I would like to express my gratitude to all the people that contributed in various ways and without whose support this thesis would probably not have been finished.

First of all, it is my pleasure to thank my advisor, Wolfgang Arendt, for his guidance and support in the last few years and his availability for discussions. Moreover, for the warm atmosphere in the group AAA created by all its members. It was no easy decision to leave Ulm in spring 2004 to go to Karlsruhe.

Therefore, I would like to thank the members of the 'Mathematisches Institut 1' of the University of Karlsruhe, for their kind reception that helped me to feel at home fast in the new environment. I am especially grateful to Lutz Weis for his interest in this thesis and his availability for questions. Thus, I have the privilege to benefit from insights and discussions with two of the three corner stones of the TULKA triangle.

I do not want to forget the whole TULKA group; the participation at the TULKA meetings is always a welcome occasion for discussions and for meeting nice people.

All the members of the two groups at Ulm and Karlsruhe helped by creating a stimulating atmosphere. In particular, I want to mention Markus Biegert with whom I began my studies in Ulm; Markus Haase, Ralph Chill and Peer Christian Kunstmann for their patience in listening to and answering my questions.

In fall 2003 I had the possibility to spend several weeks at the Tsinghua University in Beijing; I am indebted to Shangquan Bu for his hospitality and the warm welcome. It was a very exciting and enriching experience. I also acknowledge gratefully the possibility of several short stays at the University of Besançon during which I benefitted from interesting discussions with Christian Le Merdy and Gilles Lancien.

The tedious task of proofreading this thesis has been facilitated by the kind help of Bernhard Haak, Markus Haase, Cornelia Kaiser, Mahamadi Warma and Jan Zimmerschied. Of course, any mistake that has not been discovered is in the responsibility of the author.

Last but not least, I appreciate the constant support and the encouragement of my parents and of my brother and sister.

Chapter 1

Functional calculus for multisectorial operators

In this chapter we will develop a holomorphic functional calculus for multisectorial operators. This is the content of Sections 2 – 5. In Section 6 we recall the notion of R-boundedness that will become important in Section 7 where we extend the calculus to operator valued functions, i.e. when we establish the Kalton-Weis theorem in the setting of multisectorial operators. In that section we also extend the functional calculus in order to encompass the class of asymptotically bisectorial operators. The operator valued functional calculus will then be applied to derive closed-sum theorems. In Section 8 we extend Dore's theorem on the boundedness of the functional calculus in real interpolation spaces to the setting of multisectorial operators. In the setting of Hilbert space, this result and results established in Section 9 give a characterization of the boundedness of the multisectorial functional calculus in terms of bounded imaginary powers. In Section 10, this characterization will be recaptured and even generalized to arbitrary Banach spaces using the sophisticated tool of square functions.

1.1   Introduction

The notion of a H^∞ functional calculus for sectorial operators was first introduced by McIntosh [90] in the Hilbert space setting and then developed in the Banach space setting by Cowling, Doust, McIntosh and Yagi [37].

Let us sketch briefly the intuition behind a functional calculus. The idea is to give meaning to the symbol $f(A)$ in a way that the mapping $f \mapsto f(A)$ satisfies some natural algebraic and continuity properties. The following construction is inspired by the Dunford-Riesz functional calculus for a bounded linear operator which is based on the Cauchy integral formula. We recall the construction of this classical calculus [34, VII.4]. Let A be a bounded operator, U an open set in $\mathbb{C}$ containing the spectrum of A and $\Gamma \subset U \setminus \sigma(A)$ is a positively oriented system of finitely many smooth curves such that $\mathbb{C} = \Gamma \cup \text{ins}(\Gamma) \cup \text{out}(\Gamma)$, $\sigma(A) \subset \text{ins}(\Gamma)$ and $\mathbb{C} \setminus U \subset \text{out}(\Gamma)$. Recall that the set $\text{ins}(\Gamma)$ is the set of all points having index 1 and the set $\text{out}(\Gamma)$ is the set of all points having index 0 with respect to the contour Γ.

In analogy with the Cauchy integral formula we set

$$f(A) = \frac{1}{2\pi i} \int_\Gamma f(z) R(z, A) \, dz.$$

It turns out that this definition is independent of the choice of the contour Γ and that the assignment $f \mapsto f(A)$ yields a homomorphism from the algebra $H(U)$ of holomorphic functions on U into the algebra of bounded linear operators on X.

Functional calculi for sectorial operators have been studied extensively in recent years. In the same way as the Cauchy integral formula gives rise to the Dunford-Riesz functional calculus, there are other functional calculi associated to other (integral) representation formulas. We just mention the Phillips functional calculus (associated to the Laplace transform), the Mellin Transform functional calculus, or the Hirsch functional calculus (see e.g. [68, Chapter XV] and [117]).

The functional calculus is a powerful tool allowing to translate identities for scalar functions into operator identities; it allows to define operators such as e^{-tA} or A^λ and to study their algebraic relationships. For example the semigroup law of analytic semigroups is thus an immediate consequence of the semigroup property of the exponential function. If the functional calculus is bounded, it provides also norm estimates. Further important applications of the functional calculus comprise the method of sums of operators and in particular L_p-maximal regularity (see [74]) or the study of (domains of) fractional powers of operators.

In this thesis we will concentrate on the Dunford-Riesz functional calculus. This calculus has been studied in detail in the sectorial case (see e.g. [2], [61], [90] and the references therein) and is considered in the bisectorial case in [14], [50] and [53]. In [64] one may find information on spectral mapping theorems for the sectorial functional calculus.

In the following we will describe the construction of a functional calculus for (N-tuples of) multisectorial operators A allowing operator valued functions f. In later sections we will apply this to obtain results on spectral projections or on the closedness of the sum of (asymptotically bi-)sectorial operators.

1.2   Multisectorial operators

We begin with some definitions. In the following X and Y are complex (nontrivial) Banach spaces. We say that a linear operator A is acting in X if its *domain* and *range*, denoted by $\mathcal{D}(A)$ and $\mathcal{R}(A)$ respectively, are subsets of X; its *spectrum* will be denoted by $\sigma(A)$ and the *resolvent set* by $\rho(A)$. We will write $\mathcal{L}(X, Y)$ to denote the algebra of all bounded linear operators from X to Y, $\mathcal{L}(X) = \mathcal{L}(X, X)$. The other notation we use is standard or defined in later parts.

Before we introduce multisectorial operators let us recall the notion of a sectorial operator.

1.2.1 Definition. For $\omega \in (0, \pi)$ and $\alpha \in [-\pi, \pi)$ we set

$$S_\omega = \{re^{is} : r > 0, s \in (-\omega, \omega)\}$$

and

$$\Sigma_{\alpha,\omega} = e^{i\alpha} S_\omega = \{re^{is} : r > 0, s \in (\alpha - \omega, \alpha + \omega)\}.$$

That is, S_ω is the open sector around $\mathbb{R}_+$ with half-opening angle equal to ω. We call α (or, more precisely, $e^{i\alpha}$) the direction of the sector $\Sigma_{\alpha,\omega}$. To cover also the case $\omega = 0$ we set $S_0 = (0, \infty)$ and $\Sigma_{\alpha,0} = e^{i\alpha}(0, \infty)$.

1.2.2 Definition. Let A be a linear operator acting on X. We call A *sectorial* if there exists $\omega \in [0, \pi)$ such that

1. $\sigma(A) \subset \overline{S_\omega}$ and

2. $M(A, \omega') := \sup\{\|\lambda R(\lambda, A)\| : \lambda \notin \overline{S_{\omega'}}\} < \infty$ for all $\omega < \omega' < \pi$.

We write $A \in \text{Sect}(\omega)$ or $A \in \text{Sect}(S_\omega)$. We will write $A \in \text{Sect}_d(\omega)$ to indicate that A is moreover densely defined and has dense range.

This notion is illustrated by Figure 1.1. A ray or line $\gamma \subset \rho(A)$ is said to be of *minimal growth* if $\sup\{\|zR(z, A)\| : z \in \gamma\} < \infty$. This is motivated by the estimate $\|R(z, A)\| \geq (\text{dist}(z, \sigma(A)))^{-1}$. We deduce from the power series expansion of the resolvent that A is sectorial if and only if $(-\infty, 0)$ is a ray of minimal growth.

We call

$$\omega(A) := \inf\{0 \leq \omega < \pi : A \in \text{Sect}(\omega)\}$$

the *spectral angle* (or *sectoriality angle*) of A.

In the following definitions we introduce (multi)sectorial operators generalizing the notions of sectorial or bisectorial operators.

1.2.3 Definition. Let $N \in \mathbb{N}$ and $\alpha = (\alpha_1, \ldots, \alpha_N) \in (-\pi, \pi]^N$ such that $\alpha_1 < \alpha_2 < \cdots < \alpha_N$. Given such α we say that $\omega = (\omega_1, \ldots, \omega_N) \in [0, \pi)^N$ is *admissible* if the closures of two sectors $\Sigma_{\alpha_k, \omega_k}$ meet only in 0. We set

$$\Sigma_{\alpha,\omega} := \bigcup_{k=1}^{N} \Sigma_{\alpha_k, \omega_k}.$$

This notion is illustrated by Figure 1.2.

Given $\omega, \omega' \in [0, \pi)^N$ we will write $\omega < \omega'$ if $\omega_k < \omega'_k$ for all $1 \leq k \leq N$. Also we write $|\omega|$ for $\omega_1 + \cdots + \omega_N$. Observe that if $\omega < \omega'$ are admissible we have $0 \leq |\omega| < |\omega'| < 2\pi$.

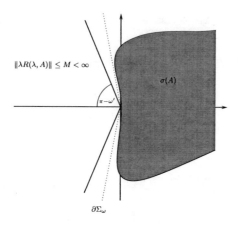

Figure 1.1: Illustration of a sectorial operator.

If α is as above and if ω is admissible we will say shortly that the pair (α, ω) is admissible.

Convention: If $N = 1$ we will identify 1-tuples and scalars.

In this case we have $S_\omega = \Sigma_\omega := \Sigma_{0,\omega}$. We obtain the bisector around the imaginary axis with half-opening angle $\theta \in (0, \pi/2)$ as $\Sigma_{\alpha,\omega}$ choosing $\alpha = (-\pi/2, \pi/2)$ and $\omega = (\theta, \theta)$. Sometimes we will drop the index α if it is clear from the context.

1.2.4 Definition. Let A be a linear operator acting in X. We call A *(multi)sectorial* if there exist $\alpha \in (-\pi, \pi]^N$ and an admissible $\omega \in [0, \pi)^N$ such that

1. $\sigma(A) \subset \overline{\Sigma_{\alpha,\omega}}$ and

2. $M(A, \alpha, \omega') := \sup\{\|\lambda R(\lambda, A)\| : \lambda \notin \overline{\Sigma_{\alpha,\omega'}}\} < \infty$ for all admissible $\omega < \omega'$.

We write $A \in \mathrm{Sect}(\alpha, \omega)$ or $A \in \mathrm{Sect}(\Sigma_{\alpha,\omega})$. We will write $A \in \mathrm{Sect}_\mathrm{d}(\alpha, \omega)$ to indicate that A is moreover densely defined and has dense range.

An operator A is in $\mathrm{Sect}(\alpha, \omega)$ if and only if each ray contained in $\mathbb{C} \setminus \overline{\Sigma_{\alpha,\omega}}$ is of minimal growth.

If $N = 2$, we will call the operator A *bisectorial*. However, if an operator is called sectorial, it will be understood in the sense of Definition 1.2.2. In view of later applications there are two important special cases of bisectorial operators which we will distinguish. We say that an operator A is *i-bisectorial* or *canonically bisectorial* if the imaginary axis is a line of minimal growth for A, i.e. if $\pm iA$ is a sectorial operator. If the real line is a line of minimal growth, we call the operator *1-bisectorial*. Clearly, A is 1-bisectorial if and only if iA is i-bisectorial.

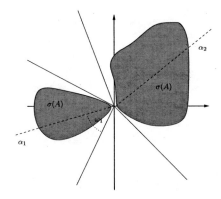

Figure 1.2: Illustration of a multisectorial operator $(N = 2)$.

We call
$$\omega_A := \inf\{0 \leq \omega < (\pi, \ldots, \pi) : A \in \text{Sect}(\alpha, \omega)\}$$
the *spectral angle* (or *sectoriality angle*) of A; here the infimum is taken in the lattice $\mathbb{R}^N$.

1.2.5 Remark. Observe that (multi)sectorial operators are always closed since the resolvent set is not empty.

We will make use of the following well-known result:

1.2.6 Lemma. *Let A be a linear operator acting in X. Assume that there is a subset Ω of $\rho(A)$ such that both 0 and ∞ are accumulation points of Ω and $\sup\{\|\lambda R(\lambda, A)\| : \lambda \in \Omega\} < \infty$. Then:*

1. $\overline{\mathcal{D}(A)} = \{x \in X : \lim_{\Omega \ni \lambda \to \infty} \lambda(\lambda - A)^{-1}x = x\}$;

2. $\overline{\mathcal{R}(A)} = \{x \in X : \lim_{\Omega \ni \lambda \to 0} x + A(\lambda - A)^{-1}x = 0\}$;

3. $\ker A \cap \overline{\mathcal{R}(A)} = \{0\}$.

Proof. See [78, Thm. 2.1, 3.1], [37, Thm. 3.8] or [61, Prop. 1.9]. □

1.2.7 Remark. If A is multisectorial then there are N intervals $I_k \subset \mathbb{R}/(2\pi\mathbb{Z})$ such that $e^{i\phi}A$ is sectorial for $\phi \in I_k$.

If we assume that A has dense range, it follows that a multisectorial operator is necessarily injective; moreover if $0 \neq z \in \rho(A)$, then $z^{-1} \in \rho(A^{-1})$ and $z^{-1}(z^{-1} - A^{-1})^{-1} = -A(z - A)^{-1}$. Therefore, if $A \in \text{Sect}(\alpha, \omega)$ we have $A^{-1} \in \text{Sect}(-\alpha, \omega)$, i.e. the spectral angle does not change.

Now we will describe the contours that will form the path of integration in order to construct the operator $f(A)$.

1.2.8 Definition. Let $\omega < \omega'$ be admissible for α. We define $\Gamma_{\omega'} = \partial \Sigma_{\alpha,\omega'} \setminus \{0\}$ oriented counterclockwise. The contour $\Gamma_{\omega'}$ is the sum of N cycles $e^{i\mu_k}\gamma_{\phi_k}$ for some $\mu_k, \phi_k \in \mathbb{R}$ where γ_ϕ is the curve parameterized by

$$\mathbb{R} \setminus \{0\} \ni t \mapsto |t|\, e^{-i\phi \operatorname{sign} t}$$

and oriented according to increasing values of t. We call $\Gamma_{\omega'}$ an admissible curve for $\Sigma_{\alpha,\omega}$.

This notion is illustrated by the Figure 1.3.

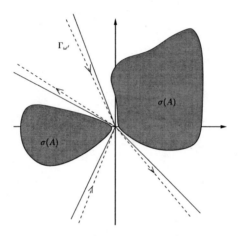

Figure 1.3: An admissible contour ($N = 2$).

1.3 H_0^∞ functional calculus for an N-tuple

In this section we will define $f(A_1, \ldots, A_N)$ for an N-tuple of multisectorial operators satisfying the following assumptions: For $N \in \mathbb{N}$ and all $j, k \in \{1, \ldots, N\}$

(H1) $A_k \in \operatorname{Sect}(\alpha^k, \omega^k)$ with $\alpha^k \in (-\pi, \pi]^{N_k}$, $\omega^k \in [0, \pi)^{N_k}$ admissible for some $N_k \in \mathbb{N}$;

(H2) the resolvents of A_j and A_k commute, i.e.

$$\forall \lambda \in \rho(A_j) \forall \mu \in \rho(A_k) \quad R(\lambda, A_j)R(\mu, A_k) = R(\mu, A_k)R(\lambda, A_j);$$

(H3) let $\omega^k < \omega'^k$ be admissible. We write $\Sigma_k = \Sigma_{\alpha^k,\omega^k}$, $\Sigma'_k = \Sigma_{\alpha^k,\omega'^k}$ and $\Sigma = \prod_{k=1}^N \Sigma_k$, $\Sigma' = \prod_{k=1}^N \Sigma'_k$. Let Γ_k be an admissible curve for Σ_k contained in Σ'_k and define $\Gamma = \prod_{k=1}^N \Gamma_k$. We will then call Γ an *admissible curve* for $(A_1, \ldots, A_N)$.

Convention: If in the following sections an N-tuple $\mathbf{A} = (A_1, \ldots, A_N)$ appears, we will always assume, if not stated otherwise, that $\mathbf{A}$ satisfies the assumptions (H1), (H2) and (H3).

Sometimes we will write $\mathbf{A}$ instead of $(A_1, \ldots, A_N)$ and we will denote by $\mathcal{B} = \mathcal{B}(\mathbf{A})$ the commutator of the set $\bigcup_{k=1}^N \{(\lambda - A_k)^{-1} : \lambda \in \rho(A_k)\}$, i.e. the closed subalgebra of the Banach algebra $\mathcal{L}(X)$ which consists of the operators that commute with the resolvents of $A_1, \ldots, A_N$.

Observe that $\{\lambda I_X : \lambda \in \mathbb{C}\} \subset \mathcal{B}$. Furthermore, we will identify complex-valued functions with functions with values in $\{\lambda I_X : \lambda \in \mathbb{C}\}$ via $(f(z))(x) = f(z)x$.

Since the contour of integration may touch the spectrum in 0 and ∞, we have to impose some growth condition on the function, more precisely we need some kind of decay. To this end we introduce the following spaces of holomorphic functions. We refer to [59] for some background on holomorphic functions of several variables and to [7] for information on vector-valued holomorphic functions.

1.3.1 Definition. Let Y be a Banach space, Ω be an open set in $\mathbb{C}^N$. We denote by

1. $H(\Omega, Y)$ the vector space of Y-valued holomorphic functions on Ω;

2. $H^\infty(\Omega, Y)$ the Banach space of Y-valued bounded holomorphic functions on Ω, equipped with the norm $\|f\|_\infty := \sup_{z \in \Omega} \|f(z)\|_Y$;

3. $H_0^\infty(\Omega, Y)$ the space of bounded holomorphic functions $f : \Omega \to Y$ *decaying regularly* in 0 and ∞, i.e. there are constants $s > 0, C > 0$ such that for all $z = (z_1, \ldots, z_N) \in \Omega$

$$\|f(z)\|_Y \leq C \prod_{j=1}^N (\min\{|z_j|, |z_j|^{-1}\})^s;$$

4. $H_P(\Omega, Y)$ the space of holomorphic functions $f : \Omega \to Y$ with polynomial growth in 0 and ∞, i.e. there are constants $s \in \mathbb{R}, C > 0$ such that for all $z = (z_1, \ldots, z_N) \in \Omega$

$$\|f(z)\|_Y \leq C \prod_{j=1}^N (\max\{|z_j|, |z_j|^{-1}\})^s.$$

If $Y = \mathbb{C}$ we will usually not mention Y.

1.3.2 Remarks.   1. Clearly, the following inclusions hold:

$$H_0^\infty(\Omega, Y) \subset H^\infty(\Omega, Y) \subset H_P(\Omega, Y) \subset H(\Omega, Y).$$

2. We say that $f \in H(\Omega, Y)$ *decays regularly at* 0 if $f(z) = O(|z|^s)$ for $z \to 0$ and some $s > 0$. Analogously, we say that f *decays regularly at* ∞ if $f(z) = O(|z|^s)$ for $z \to \infty$ and some $s < 0$.

3. If Y is a Banach algebra, then so is $H^\infty(\Omega, Y)$, and $H_0^\infty(\Omega, Y)$ is a two-sided ideal of $H^\infty(\Omega, Y)$. In addition, $H_P(\Omega, Y)$ is an algebra.

1.3.3 Theorem. *Let* $f \in H_0^\infty(\Sigma', \mathcal{B})$. *Then the integral*

$$\int_\Gamma f(z) \prod_{k=1}^N (z_k - A_k)^{-1} \, dz$$

converges in the norm of $\mathcal{L}(X)$. *Its value is independent of the choice of* Γ.

Proof. As f decays regularly in 0 and ∞ and as A_k is multisectorial there are constants $C, s > 0$ such that

$$\left\| f(z) \prod_{k=1}^N (z_k - A_k)^{-1} \right\| \leq C \prod_{k=1}^N |z_k|^{-1} \prod_{k=1}^N (\min\{|z_k|, |z_k|^{-1}\})^s = C \prod_{k=1}^N \min\{|z_j|^{s-1}, |z_j|^{-s-1}\}$$

for $z \in \Gamma$. This proves integrability. Combining this estimate with the Cauchy integral theorem gives (as in the well-known case of a sectorial operator) the independence of the integral from the choice of the contour Γ. $\qquad\square$

Now we are able to define $f(\mathbf{A})$:

1.3.4 Definition. Let $f \in H_0^\infty(\Sigma', \mathcal{B})$. We define:

$$f(\mathbf{A}) = (2\pi i)^{-N} \int_\Gamma f(z) \prod_{k=1}^N (z_k - A_k)^{-1} \, dz.$$

1.3.5 Remarks.   1. If $f \in H_0^\infty(\Sigma', \mathcal{B})$ we have the estimate $\|f(\mathbf{A})\| \leq c \int_\Gamma \|f(z)\| \frac{|dz|}{|z|}$ for some $c > 0$.

2. Let $N = 1$. If $A \in \mathrm{Sect}(\alpha, \omega) \cap \mathrm{Sect}(\alpha', \omega')$ and $f \in H_0^\infty(\Sigma_{\alpha, \omega + \epsilon 1} \cup \Sigma_{\alpha', \omega' + \epsilon 1})$ for some ϵ small enough, then it is a consequence of Cauchy's theorem that the definition above is independent of the chosen contour, i.e. we can view A as an element in $\mathrm{Sect}(\alpha, \omega)$ or $\mathrm{Sect}(\alpha', \omega')$. In particular, if $A \in \mathrm{Sect}(\alpha, \omega)$ is sectorial or bisectorial the above definition is consistent. These considerations remain also valid for $N > 1$.

The mapping $f \mapsto f(\mathbf{A})$ preserves the multiplicative structure. Here we use decisively the fact that f takes values in the algebra $\mathcal{B}$ of resolvent commuting operators.

1.3.6 Theorem. *The mapping* $f \mapsto f(\mathbf{A})$ *is an algebra homomorphism from* $H_0^\infty(\Sigma', \mathcal{B})$ *into* $\mathcal{B}$.

Proof. Approximating the integral by Riemann sums and making use of the fact that $\mathcal{B}$ is a closed subalgebra of $\mathcal{L}(X)$ we obtain that $f(\mathbf{A}) \in \mathcal{B}$. The map is clearly linear. It remains to verify that $(fg)(\mathbf{A}) = f(\mathbf{A})g(\mathbf{A})$ whenever $f, g \in H_0^\infty(\Sigma', \mathcal{B})$. To this end we choose two admissible contours $\Gamma = \prod_{k=1}^N \Gamma_k$, $\tilde{\Gamma} = \prod_{k=1}^N \tilde{\Gamma}_k$ with the property that $\tilde{\Gamma}_k$ "lies outside of Γ_k", meaning that $\tilde{\Gamma}_k = \partial\Sigma(\alpha^k, \nu'^k) \setminus \{0\}$ and $\Gamma_k = \partial\Sigma(\alpha^k, \nu^k) \setminus \{0\}$ where $\nu < \nu'$ are admissible. Then, by the resolvent equality,

$$f(\mathbf{A})g(\mathbf{A}) = (2\pi i)^{-2N} \int_\Gamma \int_{\tilde{\Gamma}} f(z)g(w) \prod_{k=1}^N (z_k - A_k)^{-1}(w_k - A_k)^{-1} \, dw \, dz$$

$$= (2\pi i)^{-2N} \int_\Gamma \int_{\tilde{\Gamma}} f(z)g(w) \prod_{k=1}^N (w_k - z_k)^{-1}((z_k - A_k)^{-1} - (w_k - A_k)^{-1}) \, dw \, dz.$$

We expand the product and obtain a sum with 2^N summands, so that we have as many integrals, each one in the $2N$ variables $w_1, \ldots, w_N, z_1, \ldots, z_N$. By Fubini's theorem we may interchange the order of integration. By Cauchy's integral theorem we have that for all $1 \le k \le N$ and all $z \in \Gamma$

$$\int_{\tilde{\Gamma}_k} (w_k - z_k)^{-1}(w_k - A_k)^{-1} \, dw_k = 0,$$

since the function of w_k is holomorphic "outside of $\tilde{\Gamma}_k$". Hence, the $2^N - 1$ terms of the sum containing the factor $(w_k - A_k)^{-1}$ for some k vanish. Therefore,

$$f(\mathbf{A})g(\mathbf{A}) = (2\pi i)^{-2N} \int_\Gamma \int_{\tilde{\Gamma}} f(z)g(w) \prod_{k=1}^N (w_k - z_k)^{-1}(z_k - A_k)^{-1} \, dw \, dz =$$

$$(2\pi i)^{-N} \int_\Gamma f(z)g(z) \prod_{k=1}^N (z_k - A_k)^{-1} \, dz = (fg)(\mathbf{A})$$

by applying N times the residue theorem. $\qquad\square$

1.3.7 Corollary. *If $f, g \in H_0^\infty(\Sigma', \mathcal{B})$ commute pointwise, then the operator $f(\mathbf{A})$ commutes with $g(\mathbf{A})$.*

Proof. By the homomorphism property proved above we have: $f(\mathbf{A})g(\mathbf{A}) = (fg)(\mathbf{A}) = (gf)(\mathbf{A}) = g(\mathbf{A})f(\mathbf{A})$. $\qquad\square$

1.3.8 Theorem. *Assume that $N = 1, f \in H_0^\infty(\Sigma'), \lambda \in \mathbb{C} \setminus \overline{\Sigma'}$. Writing $r_\lambda(z) = (\lambda - z)^{-1}$, we have $r_\lambda f \in H_0^\infty(\Sigma')$ and $(r_\lambda f)(A) = f(A)(\lambda - A)^{-1}$.*

Proof. Using continuity and the resolvent equation we find

$$f(A)(\lambda - A)^{-1} = \int_\Gamma f(z)(z - A)^{-1}(\lambda - A)^{-1} \, dz$$

$$= \int_\Gamma \frac{f(z)}{\lambda - z}(z - A)^{-1} \, dz - \int_\Gamma \frac{f(z)}{\lambda - z} \, dz (\lambda - A)^{-1} = (r_\lambda f)(A)$$

as the second term vanishes by Cauchy's theorem. $\qquad\square$

1.3.9 Remark. The function r_λ is bounded and holomorphic on Σ'. It decays regularly in ∞; it does not decay at 0, so that at the moment we cannot insert A into r_λ. However, the function r_λ is holomorphic in 0. Thus, we can modify the contour Γ such as to avoid the origin.

In this way we can extend the map $f \mapsto f(A)$ to the class of functions f that can be written as $f = g + h$ with $g \in H_0^\infty$ and h decaying regularly in ∞ and analytic in 0 by putting $f(A) = g(A) + h(A)$. This is well-defined and one can show that this map is also a homomorphism of algebras (see [61, p.27ff]). Now, we may insert A into r_λ and we find $r_\lambda(A) = (\lambda - A)^{-1}$.

We will not pursue this approach, as in the next step we will extend the functional calculus to the much bigger class $H_P(\Sigma', \mathcal{B})$.

1.4 H^∞ functional calculus for an N-tuple

In this section will extend the functional calculus up to now defined on $H_0^\infty(\Sigma', \mathcal{B})$ to the class $H_P(\Sigma', \mathcal{B})$. Actually, we are interested in the subclass $H^\infty(\Sigma', \mathcal{B})$, but from a technical point of view the class of functions of (at most) polynomial growth is advantageous.

The extension procedure goes back to Bade [16] and McIntosh [90]. The idea is as follows. Given $f \in H_P(\Sigma', \mathcal{B})$ we choose a function $\psi \in H_0^\infty(\Sigma')$ decaying sufficiently fast such that the product ψf is itself an element of $H_0^\infty(\Sigma', \mathcal{B})$, i.e. $(\psi f)(\mathbf{A})$ is defined, and such that $\psi(\mathbf{A})$ is injective; then we put $f(\mathbf{A}) = \psi(\mathbf{A})^{-1}(\psi f)(\mathbf{A})$. We will call ψ a *regularizer* for f. For more information on this approach in a more abstract framework see [65].

From now on we will assume that $\mathbf{A}$ satisfies in addition

(H4) for all $1 \leq k \leq N$ the operator A_k is injective.

1.4.1 Definition. Let $f \in H_P(\Sigma', \mathcal{B})$ and $\psi \in H_0^\infty(\Sigma')$. We say that ψ is a regularizer for f if $\psi f \in H_0^\infty(\Sigma', \mathcal{B})$ and if $\psi(\mathbf{A})$ is injective.

In this case we define $f(\mathbf{A}) = \psi(\mathbf{A})^{-1}(\psi f)(\mathbf{A})$.

1.4.2 Proposition. *Let $f \in H_P(\Sigma', \mathcal{B})$. If defined, the operator $f(\mathbf{A})$ is closed and the definition is independent of the choice of the regularizer.*

Proof. It is clear that $f(\mathbf{A})$ is closed. In order to prove the independence let $\psi, \phi \in H_0^\infty(\Sigma')$ be two regularizers for f. Then $\psi(\mathbf{A})\phi(\mathbf{A}) = (\psi\phi)(\mathbf{A}) = (\phi\psi)(\mathbf{A}) = \phi(\mathbf{A})\psi(\mathbf{A})$, thus $\psi(\mathbf{A})^{-1}\phi(\mathbf{A})^{-1} = \phi(\mathbf{A})^{-1}\psi(\mathbf{A})^{-1}$. Now it follows that

$$\psi(\mathbf{A})^{-1}(\psi f)(\mathbf{A}) = \psi(\mathbf{A})^{-1}\phi(\mathbf{A})^{-1}\phi(\mathbf{A})(\psi f)(\mathbf{A}) = \phi(\mathbf{A})^{-1}\psi(\mathbf{A})^{-1}(\phi\psi f)(\mathbf{A}) =$$
$$\phi(\mathbf{A})^{-1}\psi(\mathbf{A})^{-1}\psi(\mathbf{A})(\phi f)(\mathbf{A}) = \phi(\mathbf{A})^{-1}(\phi f)(\mathbf{A})$$

which proves the claim. $\qquad\qquad\square$

1.4.3 Remarks.   1. It is a consequence of the homomorphism property (Theorem 1.3.6) of the functional calculus that the definition above is consistent for $f \in H_0^\infty(\Sigma', \mathcal{B})$.

2. Observe that the extension to $H_P(\Sigma', \mathcal{B})$ is essentially algebraic; it requires only the existence of a regularizer. That is, we did not make use of any kind of continuity property of the functional calculus so far.

It remains to exhibit a large enough class of regularizers. This is what we will do in the following. We begin with constructing a class of regularizing functions ψ_n for a single sectorial operator.

1.4.4 Definition. For $n \in \mathbb{N}$ we define the rational function ψ_n on $\mathbb{C} \setminus \{-n, -n^{-1}\}$ by

$$\psi_n(z) = \frac{n^2 z}{(1 + nz)(n + z)} = \frac{n^2}{n^2 - 1}\left(\frac{n}{n + z} - \frac{1}{1 + nz}\right).$$

We will write $\psi = \psi_1$.

For all $k \in \{1, \ldots, N\}$ choose η_k such that $e^{i\eta_k} \notin \overline{\Sigma_{\alpha_k, \omega'_k}}$ and $0 \in (\eta_k, \eta_k + 2\pi)$. In this case $e^{i\eta_k}(0, \infty)$ is a ray of minimal growth for A_k. Define $\psi_{k,n}$ on $\mathbb{C} \setminus \{-ne^{i\eta_k}, -n^{-1}e^{i\eta_k}\}$ by $\psi_{k,n}(z) = \psi_n(e^{-i\eta_k} z)$.

For $n, N \in \mathbb{N}$ define the rational function $\Psi_{n,N}$ on $\prod_{k=1}^{N}(\mathbb{C} \setminus \{-ne^{i\eta_k}, -n^{-1}e^{i\eta_k}\})$ by $\Psi_{n,N}(z_1, \ldots, z_n) = \prod_{k=1}^{N} \psi_{k,n}(z_k)$. If N is known from the context we will just write $\Psi = \Psi_{1,N}$. Similarly we will write $\Psi_n = \Psi_{n,1}$.

It is easy to see that the functions ψ_n are decaying regularly both in 0 and in ∞. The next lemma (whose easy proof can be found in [50, Lemma 2.7]) gives a precise estimate.

1.4.5 Lemma. Let $n \in \mathbb{N}$, $\theta \in (0, \pi)$. Then we have for all $z \in S_\theta$

$$|\psi_n(z)| \leq (\cos(\theta/2))^2 \min\{1, n|z|, n|z|^{-1}\}.$$

And as a consequence we obtain:

1.4.6 Corollary. $\Psi_{n,N} \in H_0^\infty(\Sigma')$.

1.4.7 Remark. If A is invertible or f has a holomorphic extension to a neighborhood about zero, the regularizing function does not have to compensate for the growth of f or the resolvent in zero. In this case one could take a regularizer based on the function $\phi_n(z) = n(n + z)^{-1}$.

Given $f \in H_P(\Sigma', \mathcal{B})$ it is easy to see that $\Psi_{n,N}^m f \in H_0^\infty(\Sigma', \mathcal{B})$ for $m \in \mathbb{N}$ sufficiently large. The next theorem shows that $\Psi_{n,N}^m$ is a regularizer for f and that the functional calculus constructed so far is indeed related to **A** and non-trivial.

1.4.8 Theorem. Let $n, N \in \mathbb{N}$, then

$$\Psi_{n,N}(\mathbf{A}) = \prod_{k=1}^{N} e^{i\eta_k} n A_k (e^{i\eta_k} n^{-1} + A_k)^{-1} (e^{i\eta_k} n + A_k)^{-1}.$$

Proof. As the growth condition clearly implies absolute integrability, we may interchange the order of integration by Fubini's theorem, thus $\Psi_{n,N}(\mathbf{A}) = \prod_{k=1}^{N} \psi_{k,n}(A_k)$. Therefore, it suffices to consider only the case $N = 1$, i.e. it remains to prove that

$$\psi_{1,n}(A) = e^{i\eta_1} n A (e^{i\eta_1} n^{-1} + A)^{-1} (e^{i\eta_1} n + A)^{-1}.$$

We omit the standard argument and refer for a proof to [50, Thm. 3.8]. □

We can look at this construction in the following way: we actually rotate each A_k such as to obtain a sectorial operator to which we apply the function ψ_n.

1.4.9 Corollary. *As* **A** *satisfies (H1)-(H4) we obtain*

$$\sup_{n \in \mathbb{N}} \|\Psi_{n,N}(\mathbf{A})\| < \infty$$

and the injectivity of $\Psi_{n,N}^m(\mathbf{A})$.

Now we will extend the domain of the functional calculus as indicated above: We obtain a regularizer as a power of the function Ψ. For reference we note:

1.4.10 Theorem. *Let* $f \in H_P(\Sigma', \mathcal{B})$, *and let* $m \in \mathbb{N}$ *be such that* $\Psi^m f \in H_0^\infty(\Sigma', \mathcal{B})$, *then* Ψ^m *is a regularizer for* f *and*

$$f(\mathbf{A}) = \Psi(\mathbf{A})^{-m} (\Psi^m f)(\mathbf{A}).$$

1.4.11 Remarks.   1. By Proposition 1.4.2 the definition is independent of the choice of $m \in \mathbb{N}$.

2. Let $N = 1$. It is an immediate consequence of Theorem 1.3.8 that $r_\lambda(A) = R(\lambda, A)$ for $\lambda \in \mathbb{C} \setminus \overline{\Sigma'}$ and $r_\lambda(z) = (\lambda - z)^{-1}$.

3. Let $h \in H_0^\infty(\Sigma', \mathcal{B})$. Since $h(\mathbf{A})$ and $\Psi_{n,N}(\mathbf{A})$ commute by Theorem 1.3.7, it follows that for all $m \in \mathbb{N}$

$$h(\mathbf{A})(\Psi_{n,N}^m(\mathbf{A}))^{-1} \subset (\Psi_{n,N}^m(\mathbf{A}))^{-1} h(\mathbf{A})$$

(see [77, Thm. III.6.5, Problem III.5.37]). Hence, the domain of $f(\mathbf{A})$ contains the range of $\Psi^m(\mathbf{A})$.

4. Further properties of functional calculi (for a single sectorial operator) can be found in [61], [63], [81] or [117].

We collect some properties of this functional calculus:

1.4.12 Theorem.   *1. If* $f \in H_P(\Sigma', \mathcal{B})$ *and* $\lambda \in \mathbb{C} \setminus \{0\}$, *then* $\lambda f(\mathbf{A}) = (\lambda f)(\mathbf{A})$.

2. If $f : \Sigma' \to \mathcal{B}$ *is constant with unique value* $S \in \mathcal{B}$, *then* $f(\mathbf{A}) = S$.

3. If $f, g \in H_P(\Sigma', \mathcal{B})$, then $f(\mathbf{A}) + g(\mathbf{A}) \subset (f + g)(\mathbf{A})$ and $f(\mathbf{A})g(\mathbf{A}) \subset (fg)(\mathbf{A})$ with
$\mathcal{D}(f(\mathbf{A})g(\mathbf{A})) = \mathcal{D}((fg)(\mathbf{A})) \cap \mathcal{D}(g(\mathbf{A}))$.

4. If $f, g \in H_P(\Sigma', \mathcal{B})$ such that $fg = I$, then $f(\mathbf{A})$ is injective with $(f(\mathbf{A}))^{-1} = g(\mathbf{A})$.

Proof. The first two statements are easy to see. The proof of the third one does not depend on the explicit construction of the functional calculus but only on its algebraic properties. The fourth assertion follows readily from the third one. The proofs are done as in the well-known case of a single sectorial operator (with scalar-valued functions), see [90, p.7], [50, Section 4], [63, Chapter 1] or [81, Section 15B]. □

1.4.13 Corollary. *Let $f, g \in H_P(\Sigma', \mathcal{B})$.*

1. If $g(\mathbf{A}) \in \mathcal{L}(X)$, then $f(\mathbf{A}) + g(\mathbf{A}) = (f + g)(\mathbf{A})$.

2. If $g(\mathbf{A}) \in \mathcal{L}(X)$, then $f(\mathbf{A})g(\mathbf{A}) = (fg)(\mathbf{A})$.

Approximation In the following we will consider some approximation results such as the *Convergence Lemma*. For this purpose we will assume in addition that

(H5) each operator A_k is densely defined and has dense range.

We will say abusively that $\mathbf{A}$ has *dense domain and range*. Observe that, if A_k has dense range, it is automatically injective, i.e. condition (H5) implies (H4).

Note also that in the case of a single operator A this assumption is not very restrictive as we can always consider the *injective part*, i.e. the part of A in $\overline{\mathcal{R}(A)}$, which is a multisectorial operator with dense range.

Observe that $\Psi_{n,N}^m(\mathbf{A})$ converges strongly to the identity operator. This is an immediate consequence of Lemma 1.2.6 and the following continuity property of multiplication with respect to the strong operator topology.

1.4.14 Lemma. *Let $(S_\iota)_\iota$, $(T_\iota)_\iota$ be two bounded nets in $\mathcal{L}(X)$ converging strongly to S and T respectively. Then $S_\iota T_\iota$ converges strongly to ST.*

Proof. For $x \in X$ we have $\|(S_\iota T_\iota - ST)x\| \leq \|(S_\iota - S)Tx\| + \|S_\iota(T_\iota - T)x\|$, which proves the result. □

As in the case of a single sectorial operator we have:

1.4.15 Theorem. *Let $m \in \mathbb{N}$. Let $\mathbf{A}$ be densely defined with dense range then the following holds.*

1. The range of $\Psi_{n,N}(\mathbf{A})^m$ is independent of n and hence dense in X.

2. For $N = 1$, $\mathcal{R}(\psi(A)^m) = \mathcal{D}(A^m) \cap \mathcal{R}(A^m)$.

Proof. See [50, Thm. 3.10]. The argument for $N = 1$ is well known (see e.g. [61], [81]). If $N > 1$ the result follows using additionally the fact that the resolvents of A_k commute. □

In particular, the domain of $f(\mathbf{A})$ is dense in this case. The next theorems give some information on approximation. The first result we prove is the Convergence Lemma. It is a continuity property of the functional calculus. For its proof we need the following result.

1.4.16 Lemma. *Let $A \in \mathrm{Sect}(\alpha, \omega)$ and let $(f_\iota)_\iota \subset H_0^\infty(\Sigma_{\alpha,\omega'}, \mathcal{B})$, for $\omega < \omega'$, be a net of functions converging pointwise to a function f. Assume that there are constants $C, s > 0$ such that*

$$\|f_\iota(z)\| \leq C \prod_{j=1}^N (\min\{|z_j|, |z_j|^{-1}\})^s \tag{1.1}$$

for each ι and each $z \in \Sigma_{\alpha,\omega'}$. Then $f \in H_0^\infty(\Sigma_{\alpha,\omega'}, \mathcal{B})$ and $\|f_\iota(\mathbf{A}) - f(\mathbf{A})\| \to 0$.

If the net $(f_\iota)_\iota$ converges pointwise in the strong operator topology, then $f_\iota(A) \to f(A)$ strongly.

Proof. We deduce from Cauchy's theorem that $(f_\iota)_\iota$ converges uniformly on compact sets and that f is holomorphic. The estimate (1.1) also holds for the limit, consequently f is in $H_0^\infty(\Sigma_{\alpha,\omega'}, \mathcal{B})$. As $f_\iota \to f$ uniformly on compact sets, the claim follows from a version of Lebesgue's theorem. □

If we do not consider a net but only a sequence, we only require pointwise convergence in the argument above.

1.4.17 Proposition (Convergence Lemma). *Let $\mathbf{A} \in \mathrm{Sect}(\alpha, \omega)$ and $(f_\iota)_\iota \subset H^\infty(\Sigma', \mathcal{B})$ be a uniformly bounded net. Assume that the limit $f(z) = \lim_\iota f_\iota(z)$ exists pointwise on Σ' (in the strong operator topology). Then*

$$f_\iota(\mathbf{A})x \to f(\mathbf{A})x$$

for all $x \in \mathcal{R}(\Psi(\mathbf{A}))$.

Moreover, if $\mathbf{A}$ is densely defined with dense range and $\sup_\iota \|f_\iota(\mathbf{A})\| < +\infty$, then $f(\mathbf{A})$ is bounded and $f_\iota(\mathbf{A}) \to f(\mathbf{A})$ strongly.

Proof. By Vitali's theorem we know $f \in H^\infty(\Sigma', \mathcal{B})$. Define $g(z) = f(z)\Psi(z)$ and $g_\iota(z) = f_\iota(z)\Psi(z)$. The net $(g_\iota)_\iota \subset H_0^\infty(\Sigma', \mathcal{B})$ satisfies the hypotheses of Lemma 1.4.16. Hence, $g_\iota(\mathbf{A}) = f_\iota(\mathbf{A})\Psi(\mathbf{A}) \to g(\mathbf{A}) = f(\mathbf{A})\Psi(\mathbf{A})$ strongly. Therefore,

$$f_\iota(\mathbf{A})x \to f(\mathbf{A})x$$

for all $x \in \mathcal{R}(\Psi(\mathbf{A}))$.

If $\sup_\iota \|f_\iota(\mathbf{A})\| = C < +\infty$ and $\Psi(\mathbf{A})$ has dense range, this limit exists for all $x \in X$ and defines a bounded operator T with $\|T\| \leq C$, by a 3ϵ-argument. Since $f(\mathbf{A}) \subset T$ is closed, we have $\mathcal{D}(f(\mathbf{A})) = X$. □

Specializing, we immediately obtain the following corollary as a consequence of the Convergence Lemma.

1.4.18 Corollary. *Let $f \in H^\infty(\Sigma', \mathcal{B})$ and assume that $\mathbf{A}$ has dense domain and dense range. A necessary and sufficient condition for the operator $f(\mathbf{A})$ to be bounded is that $\sup_{n \in \mathbb{N}} \|(\Psi_{n,N} f)(\mathbf{A})\| < \infty$. In this case we have for all $x \in X$*

$$f(\mathbf{A})x = \lim_{n \to +\infty} (\Psi_{n,N} f)(\mathbf{A})x.$$

1.4.19 Remark. Let $f \in H_P(\Sigma', \mathcal{B})$ and $m \in \mathbb{N}$ such that $\Psi^m f \in H_0^\infty(\Sigma', \mathcal{B})$. Then $\mathcal{R}(\Psi^m(\mathbf{A}))$ is a core for $f(\mathbf{A})$ and we can characterize the domain of $f(\mathbf{A})$ by means of an approximation:

$$\mathcal{D}(f(\mathbf{A})) = \{x \in X : \lim_{n \to \infty} (\Psi_{n,N}^m f)(\mathbf{A})x \text{ exists in } X\}$$

and in this case

$$f(\mathbf{A})x = \lim_{n \to \infty} (\Psi_{n,N}^m f)(\mathbf{A})x.$$

For a proof (in the case of a single operator and a scalar-valued function f) we refer to [81, Thm. 15.8].

1.4.20 Theorem. *Let $\mathcal{A}$ be a closed subalgebra of $\mathcal{B}$ and assume that $\mathbf{A}$ has dense domain and range. The following statements are equivalent:*

1. *For all $f \in H^\infty(\Sigma', \mathcal{A})$ we have $f(\mathbf{A}) \in \mathcal{L}(X)$;*

2. *There exists $C > 0$ such that for all $f \in H_0^\infty(\Sigma', \mathcal{A}) : \|f(\mathbf{A})\| \le C \|f\|_\infty$.*

Proof. The implication $1 \Rightarrow 2$ is an immediate consequence of the closed graph theorem; the closedness of the map $f \mapsto f(\mathbf{A})$ is a consequence of the Convergence Lemma. The converse follows after an application of the previous theorem. Indeed, let $f \in H^\infty(\Sigma', \mathcal{A})$, then $\Psi_{n,N} f \in H_0^\infty(\Sigma', \mathcal{A})$ with

$$\|(\Psi_{n,N} f)(\mathbf{A})\| \le C \|\Psi_{n,N} f\|_\infty \le C' \|f\|_\infty.$$

Therefore, $f(\mathbf{A})$ is bounded. $\qquad\square$

We remark that there are sectorial operators that do not satisfy these equivalent conditions, even in Hilbert space there are counterexamples (see e.g. Example 2.2.5). Therefore, we distinguish those operators that satisfy them.

1.4.21 Definition. If condition (2) of the theorem above holds true, we say that $\mathbf{A}$ has a *bounded $H^\infty(\Sigma', \mathcal{A})$-functional calculus*. Sometimes we will write shortly $A \in H^\infty(\Sigma', \mathcal{A})$, or $A \in H^\infty$ if the sets Σ' and $\mathcal{A}$ are clear from the context.

Similar terminology is used for other types of operators and their corresponding functional calculi.

1.4.22 Remark. Observe that even if $\mathbf{A}$ has a bounded $H^\infty(\Sigma')$-functional calculus, its $H^\infty(\Sigma', \mathcal{A})$-functional calculus may very well be unbounded [83, Section 6].

As in the sectorial case we define for a single operator $A \in \mathrm{Sect}(\alpha, \omega) \cap H^\infty$ the angle

$$\omega_{H^\infty}(A) = \inf\{\omega < \phi : A \in H^\infty(\Sigma_{\alpha, \phi}, \mathcal{A})\}.$$

It is easy to see that, if $A \in H^\infty(\Sigma', \mathcal{A})$ and if $\Sigma' \leq \Sigma''$, then $A \in H^\infty(\Sigma'', \mathcal{A})$. The converse is in general not true. However, it is possible to truncate the domain in the following two cases.

1.4.23 Lemma. *Let $A \in \mathrm{Sect}(\alpha, \omega)$ be injective with $A \in H^\infty(\Sigma_{\alpha, \omega'}, \mathcal{A})$.*

1. *If $A \in \mathcal{L}(X)$, then for each $r > r(A)$ the $H^\infty(\Sigma_{\alpha, \omega'} \cap B_r(0), \mathcal{A})$-functional calculus is bounded.*

2. *If A is invertible, then for each $r > r(A^{-1})$ the $H^\infty(\Sigma_{\alpha, \omega'} \setminus \overline{B_{r^{-1}}(0)}, \mathcal{A})$-functional calculus is bounded.*

Proof. A proof of the assertion in the sectorial case can be found in [64]. □

If A is an operator in X and $c \in \mathbb{C} \setminus \{0\}$, then A and cA have the same domain and range, $\rho(cA) = c\rho(A)$ with

$$(c\lambda - cA)^{-1} = c^{-1}(\lambda - A)^{-1}.$$

Hence, if $c = re^{is}$ and $A \in \mathrm{Sect}(\alpha, \omega)$, then $cA \in \mathrm{Sect}(\alpha + s\mathbf{1}, \omega)$. In particular, if $c_1, \ldots, c_N \in \mathbb{C} \setminus \{0\}$ then $c_1 A_1, \ldots, c_N A_N$ have the same properties as $A_1, \ldots, A_N$ and their resolvents commute.

Assume $c_k = r_k e^{is_k}$ and put $\Sigma'' = \prod_{k=1}^N \Sigma_{\alpha^k + s_k \mathbf{1}, \omega'^k}$. Using this notation we have the following theorem about dilations:

1.4.24 Theorem. *Let $c_1, \ldots, c_N \in \mathbb{C} \setminus \{0\}$. If $f \in H_P(\Sigma'', \mathcal{B})$ and $g \in H_P(\Sigma', \mathcal{B})$ is defined by $g(z_1, \ldots, z_n) = f(c_1 z_1, \ldots, c_N z_N)$ then*

$$f(c_1 A_1, \ldots, c_N A_N) = g(A_1, \ldots, A_N).$$

Proof. If $f \in H_0^\infty(\Sigma'', \mathcal{B})$ the equality follows readily by a change of variables in the integral defining $f(c_1 A_1, \ldots, c_N A_N)$. Indeed, let $f \in H_0^\infty(\Sigma'', \mathcal{B})$; then g defined by $g = f \circ d$, where $d(z_1, \ldots, z_N) = (c_1 z_1, \ldots, c_N z_N)$ is in $H_0^\infty(\Sigma', \mathcal{B})$. Moreover,

$$(2\pi i)^N f(c_1 A_1, \ldots, c_N A_N) = \int_{\Gamma''} f(z) \prod_{k=1}^N R(z_k, c_k A_k) \, dz$$

$$= \int_{\Gamma'} f(d(z)) \prod_{k=1}^N R(c_k z_k, c_k A_k) c_k \, dz = \int_{\Gamma'} g(z) \prod_{k=1}^N R(z_k, A_k) \, dz = (2\pi i)^N g(A_1, \ldots, A_N),$$

where Γ'' and $\Gamma' = d^{-1}(\Gamma'')$ are suitable contours.

The general case may be done by regularization. The proof is an adaptation of [50, Thm. 4.11]. □

1.4.25 Remark. As a consequence of the theorem above, by suitable rotations, that is, multiplication by $e^{i\beta_k}$ with an appropriate β_k, one can assume all operators A_k to be sectorial. Note however that many properties of the sectorial operator obtained depend on the choice of β_k (see Example 1.9.7).

1.4.26 Theorem. *Let $f \in H_P(\Sigma'_k, \mathcal{B})$ and let $g : \Sigma' \to \mathcal{B}$ be defined by $g(z) = f(z_k)$. Then $g \in H_P(\Sigma', \mathcal{B})$ and $g(\mathbf{A}) = f(A_k)$.*

Proof. The proof is based on Fubini's theorem and regularization, cf. [50, Thm. 4.12]. $\square$

1.5 The functional calculus for a single operator

In this section we will consider (fractional) powers of a single multisectorial injective operator A. Using the same notation as before (but denoting the single operator by A instead of $\mathbf{A}$ or A_1, etc.) we assume that $e^{i\eta} \notin \overline{\Sigma'} = \overline{\Sigma_{\alpha, \omega'}}$, that is $e^{i\eta}(0, \infty)$ is a ray of minimal growth for A, where η is chosen in $(-2\pi, 0)$. In the following we will denote by $\log : \mathbb{C} \setminus e^{i\eta}(0, \infty) \to \mathbb{C}$ the logarithm with cut along the ray $e^{i\eta}(0, \infty)$. If A is (bi)sectorial we could choose the principal branch of the logarithm, that is $\eta = \pi$. With the help of this logarithm we define fractional powers.

1.5.1 Definition. For $w \in \mathbb{C}$ we denote by p_w the function defined on $\mathbb{C} \setminus e^{i\eta}(0, \infty)$ by $p_w(z) = z^w := e^{w \log z}$.

It is easy to see that the function p_w belongs to $H_P(\Sigma')$. Furthermore, for $t > 0$ we have $p_w(tz) = t^w p_w(z)$ and $|p_s(z)| = |z|^s$ for $s \in \mathbb{R}$. Moreover, note that $p_m = p_1^m$ for all $m \in \mathbb{Z}$. If the cut, and thus the logarithm defining the complex powers, is understood from the context, we will often write z^w instead of $p_w(z)$.

At first we state a technical lemma.

1.5.2 Lemma. *If $f, p_1 f \in H_0^\infty(\Sigma')$ then $(p_1 f)(A) = A f(A)$.*

Proof. By assumption f is integrable along Γ and by Cauchy's theorem $\int_\Gamma f(z) \, dz = 0$. As A is multisectorial, $A(z - A)^{-1}$ is bounded on Γ, and as A is closed we find

$$A f(A) = (2\pi i)^{-1} \int_\Gamma A f(z)(z - A)^{-1} \, dz = (2\pi i)^{-1} \int_\Gamma f(z)(z(z - A)^{-1} - 1) \, dz$$

$$= (2\pi i)^{-1} \int_\Gamma f(z) z(z - A)^{-1} \, dz = (p_1 f)(A).$$

$\square$

1.5.3 Theorem. *If $k \in \mathbb{Z}$, then $p_k(A) = A^k$.*

Proof. Since p_0 is constant, the case $k = 0$ is a consequence of Theorem 1.4.12(2). The case $k > 0$ is done by an argument similar to [50, Thm. 5.3]. The case $k < 0$ is then immediate consequence of Theorem 1.4.12(4). $\square$

1.5.4 Definition. For $w \in \mathbb{C}$ we set $A^w := p_w(A)$.

1.5.5 Remarks.   1. By the theorem above this definition is consistent if $w \in \mathbb{Z}$.

2. The choice of η corresponds to the choice we have in rotating A in order to obtain a sectorial operator $e^{i(\pi-\eta)}A$. Instead of defining the complex powers by means of a suitable branch of the logarithm we could equivalently assume A to be sectorial (rotating it suitably) and consider the usual complex powers defined by the principal branch of the logarithm (compare Theorem 1.4.24).

3. The example of a multiplication operator shows that this definition of the fractional powers of A does depend on the choice of η. However, we may vary η within its connected component of $\mathbb{C} \setminus \overline{\Sigma_{\alpha,\omega'}}$.

1.5.6 Theorem. Let $\lambda \in \mathbb{C} \setminus \overline{\Sigma_{\alpha,\omega}}$. Then for all $w \in \mathbb{C}$ we have $A^w(\lambda - A)^{-1} = (p_w r_\lambda)(A)$. In particular, if $0 < \mathcal{R}e(w) < 1$ and $\lambda \notin \overline{\Sigma_{\alpha,\omega'}}$ then $A^w(\lambda - A)^{-1} \in \mathcal{L}(X)$ and

$$\left\| A^w(\lambda - A)^{-1} \right\| \leq C_{w,\omega'} \left| \lambda \right|^{\mathcal{R}e(w)-1}.$$

Proof. The first statement follows from Corollary 1.4.13, the second by an estimation of the integral defining $(p_w r_\lambda)(A)$, see [50, Thm. 5.5]. $\qquad \square$

Note that the constant appearing on the right hand side actually depends only on the supremum of $\|\lambda R(\lambda, A)\|$ on the contour Γ. In particular, the constant does not change if we replace A by cA, $c > 0$.

The following *composition rule* is one of the most important properties of the functional calculus.

1.5.7 Proposition. Let $A \in \operatorname{Sect}(\alpha, \omega)$, $\omega < \omega'$ admissible, and $g \in H_P(\Sigma_{\alpha,\omega'})$ decaying regularly in 0 such that $g(\Sigma_{\alpha,\omega'}) \subset \overline{\Sigma_{\tilde{\alpha},\tilde{\omega}}}$ and $g(A) \in \operatorname{Sect}(\Sigma_{\tilde{\alpha},\tilde{\omega}})$. Then for $f \in H_P(\Sigma_{\tilde{\alpha},\tilde{\omega}'})$, $\tilde{\omega} < \tilde{\omega}'$ admissible, we have $f \circ g \in H_P(\Sigma_{\alpha,\omega'})$ with

$$(f \circ g)(A) = f(g(A)).$$

Proof. The proof is analogous to the sectorial case, see for example [61, Prop. 1.15], [65] or [81, Prop. 15.11]. $\qquad \square$

Another tool that is frequently quite useful is the *scaling property*. We assume that the operator A is sectorial, i.e. $\eta = -\pi$.

1.5.8 Theorem. Let $A \in \operatorname{Sect}(\alpha, \omega)$ with $\eta = -\pi$. Then, for $0 < w < 1$, we have $A^w \in \operatorname{Sect}(w\alpha, w\omega)$.

Proof. It suffices to adapt the proof given in [61, Prop. 2.2] for the case of a sectorial operator A. $\qquad \square$

We conclude this section with some remarks on the *uniqueness* of the functional calculus and by considering some examples.

Uniqueness In the following we will assume that the sectorial operator A is densely defined with dense range. We can think of the functional calculus constructed above as a (possibly) unbounded operator Φ_A from $H^\infty(\Sigma')$ into $\mathcal{L}(X)$ with domain $H_0^\infty(\Sigma')$. This operator has a closed extension which we denote by $\bar{\Phi}_A$ with domain $H_A^\infty(\Sigma')$. Now, if $\Phi : H_A^\infty(\Sigma') \to \mathcal{L}(X)$ is another map with the same properties as the functional calculus, that is, it is an homomorphism of algebras, assigns to r_λ the resolvent operator $R(\lambda, A)$ and satisfies the continuity property formulated in the convergence lemma (Prop. 1.4.17), then $\Phi = \bar{\Phi}_A$. For details we refer to [81, 9.7] or [61, 1.37]. We remark that if A has a bounded $H^\infty(\Sigma')$-functional calculus, then the map $\bar{\Phi}_A : H^\infty(\Sigma') \to \mathcal{L}(X)$, $f \mapsto f(A)$ is a closed extension of Φ_A.

1.5.9 Example. If $A \in \text{Sect}(\alpha, \omega)$ is a multiplication operator on L_p, say $Ag = mg$, then we can define $f(A)$ for $f \in H^\infty(\Sigma_{\alpha,\omega'})$ by setting $f(A)g = f(m)g$. It follows from the foregoing remarks that this map gives the bounded $H^\infty(\Sigma_{\alpha,\omega'})$-functional calculus. Actually, this definition extends to $f \in C(\Sigma_{\alpha,\omega})$.

The same is true if the operator A is essentially a multiplication operator. Let X be a UMD Banach space and $p \in (1, \infty)$. For example, the operator $A = \frac{d}{dt}$ defined on $L^p(\mathbb{R}, X)$ with domain $W^{1,p}(\mathbb{R}, X)$ is a Fourier multiplier with symbol $m(x) = ix$. In fact, for $g \in \mathcal{S}(\mathbb{R}, X)$ we have

$$Ag = \mathcal{F}^{-1}[m\mathcal{F}g], \tag{1.2}$$

where $\mathcal{F}$ denotes the Fourier transform. It is a consequence of the Mikhlin-Weis multiplier theorem [121, 122] that, given $f \in H^\infty(\Sigma_{\alpha,\omega'})$, where $\alpha = (-\pi/2, \pi/2)$ and $\omega' > 0$, and $g \in \mathcal{S}(\mathbb{R}, X)$

$$f(A)g = \mathcal{F}^{-1}[f(m)\mathcal{F}g]$$

defines the bounded $H^\infty(\Sigma_{\alpha,\omega'})$-functional calculus of A. This result can also be deduced from the fact the A generates a bounded group on a UMD space [67].

We just have seen that the derivative is a special instance of a *Fourier multiplier* (on $L_p(\mathbb{R}, X)$); the function m associated to the multiplier by (1.2) is called its *symbol*. For more information on Fourier multipliers see [42], [69], [122] or [9] (for the discrete case).

1.5.10 Remark. The notion of a UMD Banach space was introduced by Burkholder. A Banach space X is a UMD space if and only if the Hilbert transform defined on the Schwartz space extends to a bounded linear operator on $L_p(\mathbb{R}, X)$, for one (for all) $1 < p < \infty$. For some background we refer to [19, 20, 24, 106]. UMD refers to the fact that X is a UMD space if and only if all martingale difference sequences induce an unconditional decomposition of the space $L_p(S, X)$, where (S, μ) is a probability space [124, Section 1.3]. Examples of UMD Banach spaces are the reflexive $L_p(\Omega)$, Sobolev spaces $W^{s,p}(\Omega)$, Besov spaces $B_{pq}^s(\Omega)$, $p, q \in (1, \infty), s \in \mathbb{R}$, and their closed subspaces. Moreover, if X is a UMD space, then so is $L_p(\Omega, X)$ for $p \in (1, \infty)$.

1.5.11 Example. There are some classes of sectorial operators that are well known to admit a bounded H^∞-functional calculus, for example

1. normal operators in Hilbert space;

2. m-accretive operators in Hilbert space;

3. generators of bounded C_0-groups on L_p-spaces, $1 < p < \infty$;

4. negative generators of positive contraction semigroups on L_p-spaces, $1 < p < \infty$.

By the spectral theorem a normal operator is similar to a multiplication operator which implies 1. (see Example 1.5.9). The case of an m-accretive operator is reduced by Nagy's dilation theorem to the case of a unitary group, which is covered under 1. (see [61, 4.9], [6, Section 5.2] or [81, Thm. 11.5] for a direct argument). The classes 3. and 4. are obtained by the Coifman-Weiss transference principle (see [67], for 4. see also [36]). We remark that in 3. and 4. it is possible to admit vector-valued spaces $L_p(\Omega, Y)$ for a UMD space Y.

1.6   R-boundedness

Let X and Y be Banach spaces. A set of linear operators $\tau \subset \mathcal{L}(X, Y)$ is called *R-bounded* if there is a $p \in [1, \infty)$ and a constant $c_p \in [0, \infty)$ such that, for all $T_1, \ldots, T_m \in \tau$ and $x_1, \ldots, x_m \in X$, $m \in \mathbb{N}$ we have

$$\left\| \sum_{n=1}^{m} r_n T_n x_n \right\|_{L_p([0,1],Y)} \leq c_p \left\| \sum_{n=1}^{m} r_n x_n \right\|_{L_p([0,1],X)}$$

where $r_n(t) = \operatorname{sign} \sin(t2^n \pi)$ are the Rademacher functions on $[0, 1]$. We denote the infimum of these possible constants c_p by $\mathcal{R}_p(\tau)$. If the set is not R-bounded, we put $\mathcal{R}_p(\tau) = \infty$ for all p.

The Rademacher functions could be replaced by some other sequence of independent identically $\{\pm 1\}$-distributed Bernoulli random variables. Observe that by the Khinchine-Kahane inequalities the definition above does not depend on p; equivalently we could consider the norm in $L_p(X) := L_p([0, 1], X)$ for $1 \leq p < \infty$ (see for example [101]).

1.6.1 Lemma. *For $p \in [1, \infty)$ there is a positive constant C_p such that for all finite sequences $(x_n) \subset X$*

$$C_p^{-1} \left\| \sum_n r_n x_n \right\|_{L_2(X)} \leq \left\| \sum_n r_n x_n \right\|_{L_p(X)} \leq C_p \left\| \sum_n r_n x_n \right\|_{L_2(X)} .$$

In view of this independence we will usually not distinguish the p; we write $\mathcal{R}(\tau)$ instead of $\mathcal{R}_2(\tau)$. We remark that in the definition we may assume that all the operators T_k are different (see [28]).

Another useful tool will be Kahane's *contraction principle* which states

1.6.2 Lemma. *For any subset $\tau \subset \mathcal{L}(X)$ and any positive constant M we have the inequality*

$$\mathcal{R}(\{\lambda T : |\lambda| \leq M, T \in \tau\}) \leq 2M\mathcal{R}(\tau). \tag{1.3}$$

Proof. For a proof we refer again to [101] or [70, Sect. 2.5]. $\qquad\square$

Every R-bounded set $\tau \subset \mathcal{L}(X,Y)$ is clearly bounded. The converse is true if and only if X has cotype 2 and Y has type 2. If $X = Y$, this requires X to be isomorphic to a Hilbert space. This result goes back to Kwapień [82], see also [9].

1.6.3 Remark. The notion of R-boundedness was introduced by Berkson and Gillespie [18]. For a presentation of further properties we refer to [28] and [81, Section 2]. In the following we will state some of the results that we will make use of; we omit proofs and refer to [28] and [81] and the references therein for more information.

Given subsets τ and σ of $\mathcal{L}(X,Y)$ and a subset ρ of $\mathcal{L}(Y,Z)$ we define

$$\sigma + \tau = \{S + T : S \in \sigma, T \in \tau\} \text{ and } \rho \circ \tau = \{R \circ T : R \in \rho, T \in \tau\}.$$

It is easy to see that then

$$\mathcal{R}(\sigma + \tau) \leq \mathcal{R}(\sigma) + \mathcal{R}(\tau) \text{ and } \mathcal{R}(\rho \circ \tau) \leq \mathcal{R}(\rho)\mathcal{R}(\tau).$$

The following convexity result will have an important corollary.

1.6.4 Theorem. *Let* $\tau \subset \mathcal{L}(X,Y)$ *be R-bounded. Then the convex hull* $\mathrm{co}(\tau)$ *and the absolute convex hull* $\mathrm{absco}(\tau)$ *of* τ *and their closures in the strong operator topology are also R-bounded and*

$$\mathcal{R}(\overline{\mathrm{co}(\tau)}^s) \leq \mathcal{R}(\tau), \qquad \mathcal{R}(\overline{\mathrm{absco}(\tau)}^s) \leq 2\mathcal{R}(\tau).$$

Proof. The main ingredient of the proof is the identity $\mathrm{co}(A \times B) = \mathrm{co}(A) \times \mathrm{co}(B)$, where A, B are subspaces of some vector space; for details we refer to [28]. $\qquad\square$

Approximating an integral by means of simple functions we obtain a very useful corollary.

1.6.5 Corollary. *Let* τ *be an R-bounded subset of* $\mathcal{L}(X,Y)$. *For every strongly measurable* $N : \Omega \to \mathcal{L}(X,Y)$ *on a* σ-*finite measure space* (Ω, μ) *with values in* τ *and every* $h \in L_1(\Omega, \mu)$ *we define an operator* $T_{N,h} \in \mathcal{L}(X,Y)$ *by*

$$T_{N,h}x = \int_\Omega h(\omega)N(\omega)x \, d\mu(\omega), \qquad x \in X.$$

Then the set $\sigma = \{T_{N,h} : \|h\|_{L_1} \leq 1, N \text{ as above}\}$ *is R-bounded and* $\mathcal{R}(\sigma) \leq 2\mathcal{R}(\tau)$.

Combining this result with the Poisson formula yields the following useful result about interpolation of R-boundedness [81, Ex. 2.16].

1.6.6 Proposition. *Let* $N \in H^\infty(\Sigma_{\theta'}, \mathcal{L}(X,Y))$ *and assume that* $\sigma = \{N(\lambda) : 0 \neq \lambda \in \partial\Sigma_\theta\}$ *is R-bounded for some* $\theta < \theta'$. *Then* $\tau = \{N(\lambda) : \lambda \in \Sigma_\theta\}$ *is R-bounded.*

1.6.7 Remarks. The notion of R-boundedness has been used by Weis to give a vector-valued analogue of the Mikhlin multiplier theorem, which allowed to characterize L_p maximal regularity of a sectorial operator by a R-boundedness condition on the resolvent (see [121, 122] and Theorem 3.1.5).

As a final remark we point out that R-boundedness is closely related to square function estimates. In fact, in L_q spaces and, more generally, in all Banach function spaces that are q-concave for some $q < \infty$ these two notions coincide (see [81, Section 2] for details; background on q-concave Banach lattices may be found in [88]).

We conclude this section introducing the notion of R-sectoriality. The definition is analogous to the definition of a sectorial operator replacing the boundedness condition by R-boundedness.

1.6.8 Definition. Let $A \in \text{Sect}(\alpha, \omega)$ and $\omega \leq \tilde{\omega}$. We say that A is *R-multisectorial* and write $A \in \text{RSect}(\alpha, \tilde{\omega})$ if $\mathcal{R}_2(\{\lambda R(\lambda, A) : \lambda \notin \overline{\Sigma_{\alpha, \omega'}}\}) < \infty$ for all admissible $\tilde{\omega} < \omega'$. The infimum of these angles $\tilde{\omega}$ is called the R-sectoriality angle of A and is denoted by $\omega_R(A)$.

1.7 An extension of the Kalton-Weis theorem

In this section we extend the functional calculus to operator valued functions having R-bounded range. This will then be applied to establish the closedness of the sum of operators. The section is divided into two subsections. First we will consider the setting of an N-tuple of commuting multisectorial operators. In the second part we will specialize considering a single bisectorial operator. However, we will allow a more complicated spectral condition in a neighborhood about the origin.

1.7.1 A vector of multisectorial operators

In this section we will again consider the case of an N-tuple $\mathbf{A}$ of multisectorial operators satisfying the assumptions (H1)-(H4). We will need the following two technical lemmas.

1.7.1 Lemma. If $t, s \in (0, \infty)$, then $\sum_{k \in \mathbb{Z}} (\min\{(2^k t)^s, (2^k t)^{-s}\}) \leq \frac{2^{s+1}}{2^s - 1}$.

Proof. See [50, Lemma 6.1]. ☐

1.7.2 Lemma. Let $s \in (0, 1)$ and $\lambda \notin \overline{\Sigma_{\alpha, \omega'}}$. Then there exists $g \in H_0^\infty(\Sigma_{\alpha, \omega'})$ such that $g(z)^2 = z^s(z - \lambda)^{-1}$ for all $z \in \Sigma_{\alpha, \omega'}$, where $z^s = p_s(z)$ is defined as in Section 1.5.

Proof. As the function $h : z \mapsto z^s(z - \lambda)^{-1}$ is holomorphic on the simply connected sets $\Sigma_{\alpha_k, \omega'_k}$ and has no zeros, we find on each sector $\Sigma_{\alpha_k, \omega'_k}$ a square root of h. Gluing these together we obtain the desired square root g. As h is in $H_0^\infty(\Sigma_{\alpha, \omega'})$ so is g:

$$|g(z)| = |h(z)|^{1/2} \leq (C \prod_{j=1}^N (\min\{|z_j|, |z_j|^{-1}\})^s)^{1/2} = C^{1/2} \prod_{j=1}^N (\min\{|z_j|, |z_j|^{-1}\})^{s/2};$$

which completes the proof. ☐

Fix for each multisectorial operator A_k, as in the case of a single operator (page p.23), a scalar η_k such that $e^{i\eta_k} \notin \overline{\Sigma_{\alpha^k,\omega'^k}}$ and $\eta_k \in (-2\pi, 0)$. As before we define a logarithm $\log_k$ corresponding to the cut $e^{i\eta_k}(0, \infty)$. By means of this logarithm we define $p_{k,w}(\xi) = \xi^w$; as before we have $p_{k,w} \in H_P(\Sigma_{\alpha^k,\omega'^k})$. To keep notation simple we will continue to write ξ^w instead of $p_{k,w}(\xi)$. Using this notation we state the following extension of [74, Prop. 4.2].

1.7.3 Theorem. *Let $f \in H_0^\infty(\Sigma', \mathcal{B})$ and $r \in (0,1)$. Then*

$$f(\mathbf{A}) = (2\pi i)^{-N} \int_\Gamma f(z) \prod_{j=1}^N z_j^{-r} A_j^r (z_j - A_j)^{-1} \, dz,$$

and the integral converges in the norm of $\mathcal{L}(X)$.

Proof. Let $A_j^r(z_j - A_j)^{-1} = h_j(\mathbf{A})$, with $h_j(\xi_1, \ldots, \xi_N) = \xi_j^r(z_j - \xi_j)^{-1}$. Then all the operators $A_j^r(z_j - A_j)^{-1}$ commute by Corollary 1.3.7. Next, for $m \in \{0, \ldots, N\}$ define

$$S_m = (2\pi i)^{-N} \int_\Gamma f(z) \prod_{j=1}^m z_j^{-r} A_j^r (z_j - A_j)^{-1} \prod_{k=m+1}^N (z_k - A_k)^{-1} \, dz.$$

The estimate obtained in Theorem 1.5.6 implies that this integral converges in the norm of $\mathcal{L}(X)$: for $z \in \Gamma$

$$\left\| f(z) \prod_{j=1}^m z_j^{-r} A_j^r (z_j - A_j)^{-1} \prod_{k=m+1}^N (z_k - A_k)^{-1} \right\|$$

$$\leq C \prod_{j=1}^N (\min\{|z_j|, |z_j|^{-1}\})^s \prod_{j=1}^N |z_j|^{-1} = C \prod_{j=1}^N \min\{|z_j|^{s-1}, |z_j|^{-1-s}\}$$

which is integrable over Γ. Since $S_0 = f(\mathbf{A})$ and as S_N is the right hand side of the inequality we want to prove, it suffices to show $S_m = S_{m-1}$ for $1 \leq m \leq N$.

To this end we fix $m \in \{1, \ldots, N\}$ and let Δ be an admissible contour contained in Σ'_m lying *on the inside* of Σ_m. Then, interchanging the order of integration yields

$$\frac{1}{2\pi i} \int_{\Gamma_m} f(z) z_m^{-r} A_m^r (z_m - A_m)^{-1} \, dz_m$$

$$= \frac{1}{2\pi i} \int_{\Gamma_m} f(z) z_m^{-r} \frac{1}{2\pi i} \int_\Delta \frac{\xi^r}{z_m - \xi} (\xi - A_m)^{-1} \, d\xi \, dz_m$$

$$= \frac{1}{2\pi i} \int_\Delta \xi^r \frac{1}{2\pi i} \int_{\Gamma_m} \frac{z_m^{-r}}{z_m - \xi} f(z) \, dz_m (\xi - A_m)^{-1} \, d\xi$$

$$= \frac{1}{2\pi i} \int_\Delta f(z_1, \ldots, z_{m-1}, \xi, z_{m+1}, \ldots, z_N)(\xi - A_m)^{-1} \, d\xi$$

$$= \frac{1}{2\pi i} \int_{\Gamma_m} f(z)(z_m - A_m)^{-1} \, dz,$$

which proves the claim. Note, both z_m^{-r} and ξ^r are defined by means of the logarithm function $\log_m$. The application of Fubini's theorem is justified by the following estimate.

$$\int_{\Gamma_m} \int_\Delta \left\| f(z) \frac{z_m \xi^{r-1}}{z_m - \xi} \right\| |d\xi| |dz_m| = \int_{\Gamma_m} \int_\Delta \left\| f(z) z_m^{-1} \right\| \left| \frac{\zeta^{r-1}}{\frac{z_m}{|z_m|} - \zeta} \right| |d\zeta| |dz_m|$$

which is finite since $\{z_m / |z_m| : z_m \in \Gamma_m\}$ is a finite set disjoint from Δ. $\qquad\square$

The next lemma relates the notions of a (bounded) functional calculus and unconditionality. It will be a key-ingredient in the proof of the main result of this section. Observe the likeness to the square functions to be considered in Section 1.10.

1.7.4 Lemma. *Assume that* $\mathbf{A}$ *has a bounded functional calculus. Then for all* $f \in H_0^\infty(\Sigma')$ *there is a constant* $C > 0$ *with the property that for any family* $(a_k)_{k \in \mathbb{Z}^N} \subset \mathbb{C}$ *such that only finitely many* a_k *are non-zero, and* $(t_1, \ldots, t_N) \in (0, \infty)^N$, *we have:*

$$\left\| \sum_{k \in \mathbb{Z}^N} a_k f(2^{k_1} t_1 A_1, \ldots, 2^{k_N} t_N A_N) \right\| \leq C \max_{k \in \mathbb{Z}^N} |a_k|.$$

Proof. The proof goes along the same lines as in the sectorial case (see [74, Lemma 4.1], [50, Lemma 6.4] or [61, Lemma 1.39]). As $f \in H_0^\infty(\Sigma')$, there are constants $C' > 0$ and $s > 0$ such that $|f(z)| \leq C' \prod_{j=1}^N (\min\{|z_j|, |z_j|^{-1}\})^s$ for $z \in \Sigma'$. Let $(a_k)_{k \in \mathbb{Z}^N} \subset \mathbb{C}$ with $|a_k| \leq 1$ and let $t_j > 0$. Then, by Theorem 1.4.24 and the cone structure of Σ'

$$\left\| \sum_{k \in \mathbb{Z}^N} a_k f(2^{k_1} t_1 A_1, \ldots, 2^{k_N} t_N A_N) \right\| = \left\| \left(\sum_{k \in \mathbb{Z}^N} a_k f(2^{k_1} t_1 \cdot, \ldots, 2^{k_N} t_N \cdot) \right)(A) \right\|$$

$$\leq c \sup_{z \in \Sigma'} \left| \sum_k a_k f(2^{k_1} t_1 z_1, \ldots, 2^{k_N} t_N z_N) \right| \leq c C' \sup_{z \in \Sigma'} \sum_k \prod_{j=1}^N (\min\{|2^{k_j} t_j z_j|, |2^{k_j} t_j z_j|^{-1}\})^s$$

$$= c C' \sup_{z \in \Sigma'} \prod_{j=1}^N \sum_k (\min\{|2^{k_j} t_j z_j|, |2^{k_j} t_j z_j|^{-1}\})^s = c C' \sup_{t>0} \left(\sum_{j \in \mathbb{Z}} (\min\{2^j t, (2^j t)^{-1}\})^s \right)^N$$

which is a finite quantity by Lemma 1.7.1. $\qquad\square$

If $\mathbf{A}$ has a bounded scalar $H^\infty(\Sigma)$-functional calculus, its operator-valued functional calculus may be unbounded as we stated in Remark 1.4.22. However, if we consider a slightly bigger multisector and replace the boundedness condition by its 'R-analogue', we obtain a bounded functional calculus. This is the content of the following theorem which goes back to Kalton and Weis (see [74], [81, 12.7] for the sectorial case or [50] for the bisectorial case).

1.7.5 Definition. Let Ω denote an open set in $\mathbb{C}^N$. We will denote by $RH^\infty(\Omega, \mathcal{B})$ the vector space of holomorphic functions $g : \Omega \to \mathcal{B}$ with R-bounded range.

Now we will state the main theorem of this section (recall that $\Sigma \subset \Sigma'$).

1.7.6 Theorem. *Assume that* $\mathbf{A}$ *has dense domain and range and a bounded* $H^\infty(\Sigma)$ *functional calculus, that is* $f(\mathbf{A}) \in \mathcal{L}(X)$ *for all* $f \in H^\infty(\Sigma)$. *Then* $g(\mathbf{A}) \in \mathcal{L}(X)$ *for all* $g \in RH^\infty(\Sigma', \mathcal{B})$.

Moreover, there exists a constant $C = C(\mathbf{A}) > 0$ *such that for all* $g \in RH^\infty(\Sigma', \mathcal{B})$ *we have* $\|g(\mathbf{A})\| \leq C\mathcal{R}_2(g(\Sigma'))$.

Thus, the functional calculus extends to a bounded homomorphism from the algebra $RH^\infty(\Sigma', \mathcal{B})$ into the algebra of bounded linear operators $\mathcal{L}(X)$.

Proof. We fix $g \in RH^\infty(\Sigma', \mathcal{B})$ and set $g_n = \Psi_{n,N} g$ for $n \in \mathbb{N}$. We will show that $\sup_{n\in\mathbb{N}} \|g_n(\mathbf{A})\| \leq C\mathcal{R}_2(g(\Sigma'))$ which will prove the theorem by Corollary 1.4.18. Again we introduce the functions $p_{k,w}$ defined in Section 1.7 corresponding to appropriately chosen cuts $e^{i\eta_k}(0, \infty) \subset \mathbb{C} \setminus \overline{\Sigma'_k}$. Note that $g_n \in H_0^\infty(\Sigma', \mathcal{B})$; hence, by Theorem 1.7.3, we have for all $r \in (0,1)$ the following identity

$$g_n(\mathbf{A}) = (2\pi i)^{-N} \int_\Gamma g_n(z) \prod_{j=1}^{N} z_j^{-r} A_j^r (z_j - A_j)^{-1} \, dz,$$

where Γ is an admissible curve in $\Sigma' \setminus \overline{\Sigma}$ and $z_j^{-r} A_j^r$ is defined by means of the function $p_{j,r}$. The integration with respect to the variable z_j is performed on the contour Γ_j which consists of 2^{N_j} half-lines with origin at 0. Hence, it suffices to estimate each of the $\prod_{j=1}^{N} 2^{N_j}$ summands

$$(2\pi i)^{-N} \int_{\Gamma'_1} \cdots \int_{\Gamma'_N} g_n(z_1, \ldots, z_N) \prod_{j=1}^{N} z_j^{-r} A_j^r (z_j - A_j)^{-1} \, dz_N \ldots dz_1$$

where Γ'_j denotes one of the 2^{N_j} branches of Γ_j. We introduce the short hand $2^{\mathbf{k}} \mathbf{t} e^{i\beta}$ for the term $(2^{k_1} t_1 e^{i\beta_1}, \ldots, 2^{k_N} t_N e^{i\beta_N})$. If Γ'_j is the half-line passing through $e^{i\beta_j}$, where the argument β_j is chosen in $(\eta_j, \eta_j + 2\pi)$, we obtain

$$\int_{\Gamma'_1} \cdots \int_{\Gamma'_N} g_n(z_1, \ldots, z_N) \prod_{j=1}^{N} z_j^{-r} A_j^r (z_j - A_j)^{-1} \, dz_N \ldots dz_1$$

$$= \int_0^\infty \cdots \int_0^\infty g_n(\mathbf{t} e^{i\beta}) \prod_{j=1}^{N} (t_j e^{i\beta_j})^{-r} A_j^r (t_j e^{i\beta_j} - A_j)^{-1} e^{i\beta_j} \, dt_N \ldots dt_1$$

$$= \int_0^\infty \cdots \int_0^\infty g_n(\mathbf{t} e^{i\beta}) \prod_{j=1}^{N} t_j^{-r} e^{i(1-r)\beta_j} t_j^{-1} A_j^r (e^{i\beta_j} - t_j^{-1} A_j)^{-1} \, dt_N \ldots dt_1;$$

here we made use of the identities $(t_j e^{i\beta_j})^{-r} = t_j^{-r} (e^{i\beta_j})^{-r}$ and $(e^{i\beta_j})^{-r} e^{i\beta_j} = e^{i(1-r)\beta_j}$. By

Theorem 1.4.24 we have $(t_j^{-1}A_j)^r = t_j^{-r}A_j^r$, therefore the term is equal to

$$\int_0^\infty \cdots \int_0^\infty g_n(\mathbf{t}e^{i\beta}) \prod_{j=1}^N t_j^{-1} e^{i(1-r)\beta_j} (t_j^{-1}A_j)^r (e^{i\beta_j} - t_j^{-1}A_j)^{-1} \, dt_N \ldots dt_1$$

$$= \sum_{k_1 \in \mathbb{Z}} \cdots \sum_{k_N \in \mathbb{Z}} \int_{2^{k_1}}^{2^{k_1+1}} \cdots \int_{2^{k_N}}^{2^{k_N+1}} g_n(\mathbf{t}e^{i\beta}) \prod_{j=1}^N t_j^{-1} e^{i(1-r)\beta_j} (t_j^{-1}A_j)^r (e^{i\beta_j} - t_j^{-1}A_j)^{-1} \, dt_N \ldots dt_1$$

$$= \int_{[1,2]^N} \sum_{k_1 \in \mathbb{Z}} \cdots \sum_{k_N \in \mathbb{Z}} g_n(2^{\mathbf{k}}\mathbf{t}e^{i\beta}) \prod_{j=1}^N t_j^{-1} e^{i(1-r)\beta_j} (2^{-k_j}t_j^{-1}A_j)^r (e^{i\beta_j} - 2^{-k_j}t_j^{-1}A_j)^{-1} \, dt_N \ldots dt_1.$$

We applied Fubini's theorem in order to move the sums inside the integral. The required integrability of the integrand is easily seen from the inequalities that follow. Now, Theorem 1.5.6 tells us that

$$\sup_{\mathbf{t} \in [1,2]^N} \sup_{\mathbf{k} \in \mathbb{Z}^N} \left\| \prod_{j=1}^N t_j^{-1} e^{i(1-r)\beta_j} (2^{-k_j}t_j^{-1}A_j)^r (e^{i\beta_j} - 2^{-k_j}t_j^{-1}A_j)^{-1} \right\| = C(\mathbf{A}) < +\infty,$$

where the constant $C(\mathbf{A})$ depends only on the suprema of $\|z_j R(z_j, A_j)\|$ on Γ_j. Furthermore, as $g_n \in H_0^\infty(\Sigma', \mathcal{B})$, there is $s > 0$ such that for all $\mathbf{t} \in [1,2]^N$ we have

$$\left\| g_n(2^{\mathbf{k}}\mathbf{t}e^{i\beta}) \right\| \le C(g_n) \prod_{j=1}^N \min\{(2^{k_j}t_j)^s, (2^{k_j}t_j)^{-s}\}.$$

Combining this with Lemma 1.7.1 gives the estimate

$$\sup_{\mathbf{t} \in [1,2]^N} \sum_{\mathbf{k} \in \mathbb{Z}^N} \left\| g_n(2^{\mathbf{k}}\mathbf{t}e^{i\beta}) \prod_{j=1}^N t_j^{-1} e^{i(1-r)\beta_j} (2^{-k_j}t_j^{-1}A_j)^r (e^{i\beta_j} - 2^{-k_j}t_j^{-1}A_j)^{-1} \right\|$$

$$\le C(g_n)C(\mathbf{A}) \left(\frac{2^{s+1}}{2^s - 1} \right)^N,$$

which justifies the application of Fubini's theorem above.

To conclude the proof it suffices to show that the norm of

$$M_n(\mathbf{t}) = \sum_{\mathbf{k} \in \mathbb{Z}^N} g_n(2^{\mathbf{k}}\mathbf{t}e^{i\beta}) \prod_{j=1}^N t_j^{-1} e^{i(1-r)\beta_j} (2^{-k_j}t_j^{-1}A_j)^r (e^{i\beta_j} - 2^{-k_j}t_j^{-1}A_j)^{-1}$$

is bounded from above by $C\mathcal{R}_2(g(\Sigma'))$ (uniformly in $\mathbf{t}$). By Fatou's Lemma it suffices to find a norm estimate for elements of the form

$$M_{n,E}(\mathbf{t}) = \sum_{\mathbf{k} \in E} g_n(2^{\mathbf{k}}\mathbf{t}e^{i\beta}) \prod_{j=1}^N t_j^{-1} e^{i(1-r)\beta_j} (2^{-k_j}t_j^{-1}A_j)^r (e^{i\beta_j} - 2^{-k_j}t_j^{-1}A_j)^{-1}$$

for finite subsets E of $\mathbb{Z}^N$. To this end, fix $n \in \mathbb{N}$, $\mathbf{t} \in [1,2]^N$, a finite subset E of $\mathbb{Z}^N$, $x \in X$ with $\|x\| \le 1$ and $x^* \in X$ with $\|x^*\| \le 1$.

By Lemma 1.7.2 and Theorem 1.5.6 we have

$$(2^{-k_j} t_j^{-1} A_j)^r (e^{i\beta_j} - 2^{-k_j} t_j^{-1} A_j)^{-1} = h_j (2^{-k_j} t_j^{-1} A_j)^2$$

where $h_j \in H_0^\infty(\Sigma_j)$, and therefore

$$\prod_{j=1}^N (2^{-k_j} t_j^{-1} A_j)^r (e^{i\beta_j} - 2^{-k_j} t_j^{-1} A_j)^{-1} = h(2^{-k_1} t_1^{-1} A_1, \ldots, 2^{-k_N} t_N^{-1} A_N)^2$$

for $h(z) = h_1(z_1) \cdots h_N(z_N) \in H_0^\infty(\Sigma)$. Since g_n maps into $\mathcal{B}$, its values commute with $h(2^{-k_1} t_1^{-1} A_1, \ldots, 2^{-k_N} t_N^{-1} A_N)$. This yields

$$M_{n,E}(\mathbf{t}) = \sum_{\mathbf{k} \in E} \prod_{j=1}^N t_j^{-1} e^{i(1-r)\beta_j} h(2^{-k_1} t_1^{-1} A_1, \ldots, 2^{-k_N} t_N^{-1} A_N) \cdot$$
$$\cdot g_n(2^{\mathbf{k}} \mathbf{t} e^{i\beta}) h(2^{-k_1} t_1^{-1} A_1, \ldots, 2^{-k_N} t_N^{-1} A_N).$$

If $|E|$ denotes the cardinality of E, we have $\sum_{\epsilon \in \{-1,1\}^E} \epsilon_k \epsilon_l = 2^{|E|} \delta_{k,l}$; writing $2^{-\mathbf{k}} \mathbf{t}^{-1} \mathbf{A}$ for $(2^{-k_1} t_1^{-1} A_1, \ldots, 2^{-k_N} t_N^{-1} A_N)$ we obtain by randomization

$$|\langle M_{n,E}(\mathbf{t}) x, x^* \rangle| = \prod_{j=1}^N t_j^{-1} \left| \langle g_n(2^{\mathbf{k}} \mathbf{t} e^{i\beta}) h(2^{-\mathbf{k}} \mathbf{t}^{-1} \mathbf{A}) x, h(2^{-\mathbf{k}} \mathbf{t}^{-1} \mathbf{A})^* x^* \rangle \right|$$

$$\le 2^{-|E|} \left| \sum_{\epsilon \in \{-1,1\}^E} \sum_{\mathbf{k}, \mathbf{l} \in E} \langle \epsilon_k g_n(2^{\mathbf{k}} \mathbf{t} e^{i\beta}) h(2^{-\mathbf{k}} \mathbf{t}^{-1} \mathbf{A}) x, \epsilon_l h(2^{-\mathbf{k}} \mathbf{t}^{-1} \mathbf{A})^* x^* \rangle \right|$$

$$\le 2^{-|E|} \left(\sum_\epsilon \left\| \sum_{\mathbf{k} \in E} \epsilon_k g_n(2^{\mathbf{k}} \mathbf{t} e^{i\beta}) h(2^{-\mathbf{k}} \mathbf{t}^{-1} \mathbf{A}) x \right\|^2 \right)^{1/2} \left(\sum_\epsilon \left\| \sum_{\mathbf{l} \in E} \epsilon_l h(2^{-\mathbf{l}} \mathbf{t}^{-1} \mathbf{A})^* x^* \right\|^2 \right)^{1/2}.$$

Now we make use of the R-boundedness. As g has R-bounded range and as $\Psi_{n,N}$ is uniformly bounded, it follows from Kahane's contraction principle (1.3) that $\mathcal{R}_2(g_n(\Sigma')) \le C\mathcal{R}_2(g(\Sigma'))$ where C depends only on Σ'. Therefore,

$$|\langle M_{n,E}(\mathbf{t}) x, x^* \rangle|$$

$$\le 2^{-|E|} C\mathcal{R}_2(g(\Sigma')) \left(\sum_\epsilon \left\| \sum_{\mathbf{k} \in E} \epsilon_k h(2^{-\mathbf{k}} \mathbf{t}^{-1} \mathbf{A}) x \right\|^2 \right)^{1/2} \left(\sum_\epsilon \left\| \sum_{\mathbf{l} \in E} \epsilon_l h(2^{-\mathbf{l}} \mathbf{t}^{-1} \mathbf{A})^* x^* \right\|^2 \right)^{1/2}$$

$$\le 2^{-|E|} C\mathcal{R}_2(g(\Sigma')) \sum_{\epsilon \in \{-1,1\}^E} \left\| \sum_{\mathbf{k} \in E} \epsilon_k h(2^{-\mathbf{k}} \mathbf{t}^{-1} \mathbf{A}) \right\|^2.$$

Now we use Lemma 1.7.4 to estimate the norms on the right hand side. As $\mathbf{A}$ has dense domain and range and a bounded $H^\infty(\Sigma)$ functional calculus, we find that

$$\left\|\sum_{\mathbf{k}\in E} \epsilon_{\mathbf{k}} h(2^{-\mathbf{k}} \mathbf{t}^{-1} \mathbf{A})\right\|$$

is bounded by a constant depending only on h and thus only on Σ. Finally,

$$|\langle M_{n,E}(\mathbf{t})x, x^*\rangle| \leq 2^{-|E|} \sum_{\epsilon \in \{-1,1\}^E} C\mathcal{R}_2(g(\Sigma')) = C\mathcal{R}_2(g(\Sigma')),$$

which concludes the proof. $\square$

Application: A closed-sum theorem. As an application of this theorem we will give a condition assuring the closedness of the operator sum $A + B$ where A and B are two multisectorial resolvent commuting operators.

Da Prato and Grisvard [39] used first the method of sums of operators in case of sectorial operators. Recently, an analogue was established in the setting of bisectorial operators by Arendt and Bu [10].

The following theorem is an analogue of [74, Theorem 6.3]; it is in the same spirit as the well known theorem of Dore and Venni, where the operators A and B, defined on a UMD space, are assumed to admit bounded imaginary powers [48]. However, we have a stronger assumption on A, which we require to have a bounded functional calculus, but a weaker assumption on B, which is only assumed to be R-multisectorial. A special case of this theorem is the first part of [31, Thm. 6].

1.7.7 Theorem. *Let A and B be two resolvent commuting multisectorial operators having dense domain and range. Let $\Sigma = \Sigma_{\alpha,\omega}$ and $\tilde{\Sigma} = \Sigma_{\beta,\tilde{\omega}}$. Assume that*

1. $A \in H^\infty(\Sigma)$;

2. $-B \in \mathrm{RSect}(\beta, \tilde{\omega})$;

3. the closures of the sets Σ and $\tilde{\Sigma}$ meet only in zero.

Then the operator $A + B$ with domain $\mathcal{D}(A) \cap \mathcal{D}(B)$ is closed and

$$\|Ax\| + \|Bx\| \leq C \|Ax + Bx\| \tag{1.4}$$

for all $x \in \mathcal{D}(A) \cap \mathcal{D}(B)$. Moreover, the operator $A + B$ is invertible if either A or B is invertible.

Note that condition 3. corresponds to the usual (parabolicity) condition on the sum of the sectoriality angles. Given assumption 3., the spectra of A and $-B$ are disjoint if and only if either A or B is invertible. We do not have to assume that B has dense range, as we will see in Theorem 2.6.3.

Proof. The following proof is an adaptation of an analogue result for sectorial operators, see [81, Thm. 12.13] or [74, Thm. 6.3]. Let $\Sigma' = \Sigma_{\alpha,\omega'}$ where $\omega' > \omega$ is admissible and such that $\overline{\Sigma'} \cap \overline{\Sigma} = \{0\}$. Define $f : \Sigma' \to \mathcal{B}$, where $\mathcal{B}$ denotes commutator algebra of A, by $f(z) = B(z + B)^{-1}$. Then $f \in RH^\infty(\Sigma', \mathcal{B})$. As A has a bounded $H^\infty(\Sigma)$-calculus, the Kalton-Weis-theorem implies that $f(A)$ is bounded. The *intuitive reasoning* is the following: $f(A) = B(A + B)^{-1}$ is bounded which implies the inequality (1.4), and hence the closedness of $A + B$. The rigorous argument follows.

Let Ψ_n be a regularizer introduced in Definition 1.4.4 for both A and B. Define furthermore

$$g(\lambda) = (\lambda + B)\Psi_n(\lambda)^2 \Psi_n(B)^2.$$

By our assumptions on B we have $g \in RH_0^\infty(\Sigma', \mathcal{B})$. We find

$$g(\lambda)f(\lambda) = B\Psi_n(B)^2 \Psi_n(\lambda)^2$$

and hence, by the homomorphism property of the calculus,

$$(A + B)\Psi_n(A)^2 \Psi_n(B)^2 f(A) = g(A)f(A) = B\Psi_n(A)^2 \Psi_n(B)^2$$

which gives, using the boundedness of the functional calculus, for $x \in \mathcal{D}(A) \cap \mathcal{D}(B)$,

$$\left\| B\Psi_n(A)^2 \Psi_n(B)^2 x \right\| \leq \|f(A)\| \left\| (A + B)\Psi_n(A)^2 \Psi_n(B)^2 x \right\|.$$

Since $\Psi_n(A)$ and $\Psi_n(B)$ converge strongly to the identity by our density assumption, we obtain, by Lemma 1.4.14, letting n tend to infinity, the inequality

$$\|Bx\| \leq \|f(A)\| \|(A + B)x\|.$$

Therefore, $\|Ax\| + \|Bx\| \leq \|Ax + Bx\| + 2\|Bx\| \leq (2\|f(A)\| + 1)\|Ax + Bx\|$. This immediately implies the closedness of $A + B$ on $\mathcal{D}(A) \cap \mathcal{D}(B)$. Indeed, let $x_n \in \mathcal{D}(A) \cap \mathcal{D}(B)$ with $x_n \to x$ and $(A + B)x_n \to y$. By (1.4), both (Ax_n) and (Bx_n) are Cauchy sequences, hence, as both A and B are closed, we find $x \in \mathcal{D}(A) \cap \mathcal{D}(B)$ and $y = \lim_n (A + B)x_n = \lim_n Ax_n + \lim_n Bx_n = Ax + Bx$.

If A or B is invertible, we have the inequality $\|Ax\| + \|Bx\| \geq c\|x\|$, where c equals either $\|A^{-1}\|$ or $\|B^{-1}\|$. Hence, we have that

$$\|x\| \leq cC \|Ax + Bx\|,$$

which is well-known to be equivalent to the fact that $A + B$ is injective and has closed range [21, Thm. II.20]. However, the range of $A + B$ is dense in X; thus we found that $A + B$ is invertible. The density of $\mathcal{R}(A + B)$ may be deduced either from a classical result in [39] (we provide some details in Theorem 2.6.3) or, using the machinery of functional calculus, by observing that $\mathcal{R}(A + B) = \mathcal{R}((\cdot + B)(A)) = \mathcal{D}((\cdot + B)^{-1}(A))$ is dense as it contains the dense set $\mathcal{D}(A^2) \cap \mathcal{R}(A^2)$. $\qquad\square$

1.7.8 Remark. In [10] the authors consider also the case of two commuting bisectorial operators A and B defined on a Banach space X, but they allow a more complicated spectral condition in a neighborhood of the origin. Assuming that the spectra of A and $-B$ are separated they show that sum $A + B$ is closable and prove that real interpolation spaces between X and $\mathcal{D}(A)$ are maximal regularity spaces for the problem $Ay + By = x$ in X (see also [39]). In addition, they prove a spectral mapping theorem, stating that $\sigma(\overline{A + B}) \subset \sigma(A) + \sigma(B)$; in particular the operator $\overline{A + B}$ is invertible.

We will consider this setting in more detail in the following subsection on asymptotically bisectorial operators.

The theorem of Kalton and Weis implies the following R-boundedness result. In its formulation we need the notion of property (α) introduced by Pisier [100]. Given two independent $\{\pm 1\}$-distributed sequences (r_i) and $(\widetilde{r}_j)$ of Bernoulli random variables their products $(r_i \widetilde{r}_j)_{i,j}$ won't be independent in general. We say that a Banach space X has property (α) if there holds an analogue of Kahane's contraction principle for the random sequence $(r_i \widetilde{r}_j)_{i,j}$. Recall that subspaces of $L_q(\Omega)$, Sobolev spaces $W^{1,q}(\Omega)$, Besov spaces $B_{pq}^s(\Omega)$, $p, q \in (1, \infty), s \in \mathbb{R}$, and their closed subspaces have this property. For the precise definition and further background see e.g. [66, 3.1.6], [83], [81, 4.9] or [100].

We will denote by $\mathcal{A}$ the subalgebra of $\mathcal{L}(X)$ of all operators that commute with resolvents of A.

1.7.9 Theorem. *Let the Banach space X have property (α) and let $A \in \mathrm{Sect}_d(\alpha, \omega)$. For an R-bounded subset $\tau \subset \mathcal{L}(X)$ define*

$$RH^\infty(\Sigma', \tau) = \{f \in RH^\infty(\Sigma', \mathcal{A}) : f(z) \in \tau \text{ for } z \in \Sigma'\}.$$

If A has a bounded $H^\infty(\Sigma_{\alpha,\omega_1})$-functional calculus, then for $\omega_1 < \omega_2$ the set $\{f(A) : f \in RH^\infty(\Sigma_{\alpha,\omega_2}, \tau)\}$ is R-bounded.

Proof. For a proof in the sectorial case we refer to [81, 12.10]. $\qquad\square$

1.7.10 Corollary. *Under the same assumptions as in Theorem 1.7.9 we obtain in particular, if τ is the unit ball in $\mathbb{C}$, the R-boundedness of the set $\{f(A) : \|f\|_{H^\infty(\Sigma_{\alpha,\omega_2})} \leq 1\}$.*

In particular, we find that, if A has a bounded H^∞-functional calculus on a Banach space X with property (α), then it is R-sectorial and the angle of the calculus is less or equal to the R-sectoriality angle. In fact, in [74, Thm. 5.3] it is shown that it suffices for X to have property (Δ) and that the two angles agree in this case. Property (Δ) is weaker than property (α). For its definition and other background information we refer to [74] or [124].

If $\mathbf{A}$ has a bounded functional calculus, then by Theorem 1.4.26 each A_k has a bounded functional calculus. Combining Theorem 1.7.9 with Theorem 1.7.6 allows by an iteration argument to prove the converse.

1.7.11 Theorem. *Let X be a Banach space with property (α) and assume that $\mathbf{A}$ satisfies (H1)-(H5). If each A_k has a bounded $H^\infty(\Sigma_k')$-functional calculus, then for all admissible*

$\Sigma_k'' = \Sigma_{\alpha^k, \omega''^k}$ with $\omega''^k > \omega'^k$ the N-tuple $\mathbf{A}$ has a bounded $H^\infty(\Sigma'')$-functional calculus, where $\Sigma'' = \Sigma_1'' \times \cdots \times \Sigma_N''$.

Furthermore, the set of operators $\{f(\mathbf{A}) : f \in H^\infty(\Sigma''), \|f\|_{H^\infty} \leq 1\}$ is R-bounded in $\mathcal{L}(X)$.

If bounded functional calculi for the components A_i induce a bounded functional calculus for the N-tuple $\mathbf{A}$ we say that the Banach space X has the joint calculus property. The theorem states the Banach spaces with property (α) belong to that class. The Schatten spaces S_p $(2 \neq p \in [1, \infty])$ fail the joint calculus property [83, Thm. 3.9].

Proof. We proceed by induction on N. The case $N = 1$ is an immediate consequence of Corollary 1.7.10. Assume that $N \geq 2$ and that we know the theorem to be true for the $(N-1)$-tuple $\mathbf{A}' = (A_2, \ldots, A_N)$. Let Φ denote the bounded $H^\infty(\Sigma_2'' \times \cdots \times \Sigma_N'')$-functional calculus for $\mathbf{A}'$ and Ψ the operator-valued $H^\infty(\Sigma_1'', \mathcal{A}_1)$-functional calculus for A_1, where $\mathcal{A}_1$ denotes the commutator algebra of A_1. By induction hypothesis the set

$$\tau = \{g(\mathbf{A}') : g \in H^\infty(\Sigma_2'' \times \cdots \times \Sigma_N''), \|g\|_{H^\infty} \leq 1\}$$

is an R-bounded subset of $\mathcal{B} \subset \mathcal{A}_1$. Given $f \in H^\infty(\Sigma'')$ with $\|f\|_{H^\infty} \leq 1$ the set

$$\{f(z_1, \cdot, \ldots, \cdot) : z_1 \in \Sigma_1''\}$$

is uniformly bounded in $H^\infty(\Sigma_2'' \times \cdots \times \Sigma_N'')$. Hence

$$\Phi[f(z_1, \cdot, \ldots, \cdot)] = f(z_1, A_2, \ldots, A_N) \in \tau$$

for all $z_1 \in \Sigma_1''$. Furthermore, the function

$$z_1 \mapsto f(z_1, A_2, \ldots, A_N)$$

is analytic as a consequence of the theorems of Lebesgue and Vitali; therefore the function $f(\cdot, A_2, \ldots, A_N)$ is in $RH^\infty(\Sigma_1'', \mathcal{A}_1)$. Now Theorem 1.7.6 implies that $\Psi[f(\cdot, A_2, \ldots, A_N)]$ is bounded. An application of Fubini's theorem yields $\Psi[f(\cdot, A_2, \ldots, A_N)] = f(\mathbf{A})$. To conclude the proof observe that the set $\{f(\mathbf{A}) : f \in H^\infty(\Sigma''), \|f\|_{H^\infty} \leq 1\}$ is R-bounded as a consequence of Theorem 1.7.9. $\qquad\square$

1.7.2 An asymptotically bisectorial operator

Studying the periodic Cauchy problem one requires only a resolvent estimate at infinity but not in a neighborhood about the origin. Motivated by this example we introduce the notion of *asymptotically bisectorial* operators. Such operators where already considered in [10]. We will modify the arguments and notions of the preceding subsection in order to cover also this setting.

1.7.12 Definition. We will call a set Ω a *quasi-sector* if there is a bounded open set $O \subset \mathbb{C}$, a multisector $\Sigma = \Sigma_{\alpha, \omega}$ and a positive number r such that

$$\Omega = O \cup (\Sigma_{\alpha, \omega} \cap \{z : |z| > r\}).$$

We write $\Omega = \Omega_{\alpha, \omega}$ to specify the *direction* α and the *opening angle* ω of the quasi-sector.

Given two quasi-sectors Ω and Ω' we write $\Omega < \Omega'$ if $\Omega \subset \Omega'$ and $\operatorname{dist}(\Omega, \partial\Omega') > 0$.

1.7.13 Definition. Let A be a linear operator acting in the Banach space X. We will call A *asymptotically multisectorial* (asymptotically R-multisectorial) if there exists a quasi-sector $\Omega = \Omega_{\alpha,\omega}$ such that

1. $\sigma(A) \subset \Omega$;

2. the set $\{zR(z, A) : z \notin \Omega'\}$ is bounded (R-bounded) for all quasi-sectors $\Omega < \Omega'$.

We will write $A \in \mathrm{ASect}(\Omega)$ ($A \in \mathrm{ARSect}(\Omega)$) in this case. If moreover A is densely defined with dense range, we write $A \in \mathrm{ASect}_d(\Omega)$.

In the following we will restrict ourselves to the setting of *asymptotically bisectorial* operators. Given such an operator $A \in \mathrm{ASect}(\Omega_{\alpha,\omega})$ (with $\alpha \in (-\pi, \pi]^2, \omega \in [0, \pi)^2$ admissible), we will call a quasi-sector Ω *admissible* (for A) if $\Omega_{\alpha,\omega} < \Omega$.

In a first step we will define a functional calculus for operators of that type. This will be done combining the methods introduced for bisectorial operators (at infinity) and standard techniques for the Dunford-Riesz calculus (about the origin). Let us denote by $H_d^\infty(\Omega, Y)$ the set of bounded holomorphic Y-valued functions that decay regularly at infinity.

Let Ω be admissible for A; a contour Γ will be called admissible for Ω if it is a disjoint system of curves $\{\Gamma_1, \Gamma_2, \ldots, \Gamma_{k_r}\}$, where Γ_j, $j \geq 3$, are piecewise smooth Jordan curves (of finite length), and where Γ_j, $j = 1, 2$, are piecewise smooth curves (Jordan curves if considered on the Riemann sphere) with the additional property that the inside ins (Γ) is an admissible set satisfying ins $(\Gamma) \leq \Omega$ and $\mathbb{C} \setminus \Omega \subset$ out (Γ). Recall that the inside of Γ is the set of all points having index 1 with respect to that contour; the outside is the set of points of index zero. The index of the infinite contour $\Gamma_1 \oplus \Gamma_2$ is obtained by approximating it inside the domain of holomorphy Ω by Jordan curves of finite length. In addition we require that $\mathbb{C} = \Gamma \cup \mathrm{ins}\,(\Gamma) \cup \mathrm{out}\,(\Gamma)$. The situation is illustrated by the Figure 1.4

The next lemma answers positively the question concerning the existence of such admissible contours. It is the key-ingredient for the subsequent construction of an $H^\infty(\Omega)$-functional calculus for A.

1.7.14 Lemma. *Let A be an asymptotically bisectorial operator and let Ω be an admissible set. Then there exists a contour Γ which is admissible for Ω.*

Proof. The curves Γ_1 and Γ_2 are constructed using [10, Lemma 3.1] (setting there $S = \Omega$, $T = \rho(A)$ and choosing $a = b = r > 0$ sufficiently large). Having constructed these two curves the problem is reduced to the case of compact spectrum; the classical construction of the remaining Jordan curves may be found in [33, p.195]. $\qquad\square$

1.7.15 Remark. Note that the contour constructed in this way satisfies ins $(\Gamma) < \Omega$.

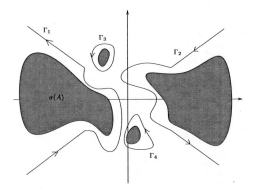

Figure 1.4: The spectrum of the asymptotically bisectorial operator A is indicated by the area hatched in grey; the inside of the contour is an admissible set for A and the contour Γ is itself admissible for its inside (assuming the proper resolvent growth).

Observe that, if Ω and Ω' are admissible for A with $\Omega < \Omega'$, then any contour admissible for Ω is also admissible for Ω'.

Given an admissible set Ω we define the operator $f(A)$ for $f \in H_d^\infty(\Omega)$ again by means of a Cauchy-type contour integral just as we did in the sectorial case

$$f(A) = \frac{1}{2\pi i} \int_\Gamma f(z) R(z, A) \, dz.$$

The integral is independent of the chosen (admissible) contour as can be seen by applying the homology version of Cauchy's theorem. The mapping $f \mapsto f(A)$ is a homomorphism of algebras. By means of the regularizer $\psi_n(z) = in(in - e^{i\theta}z)^{-1}$ (we do not have to compensate for some singularity at zero) we can extend this homomorphism to the class $H^\infty(\Omega)$. The arguments carry over from the sectorial case and this functional calculus has essentially the same properties; in particular, there is an appropriate version of the convergence lemma. Being arrived at this point one can derive a lot of the results and notions established already for the bisectorial functional calculus as the arguments mostly involve only these generic properties and do not make use of the specific construction.

1.7.16 Definition. Let A be a closed operator. If there is a complex number $w \in \mathbb{C}$ such that $A + w$ is a sectorial (R-sectorial) operator, we will call A quasi-sectorial (quasi-R-sectorial). We write $A \in \mathrm{QSect}(\Sigma)$ ($A \in \mathrm{QRSect}(\Sigma)$) if $A + w \in \mathrm{Sect}(\Sigma)$ ($\mathrm{RSect}(\Sigma)$). If moreover A has dense domain and range, we write $A \in \mathrm{QSect}_d(\Sigma)$. If $A + w$ has a bounded $H^\infty(\Sigma, \mathcal{A})$-functional calculus, we will write similarly $A \in QH^\infty(\Sigma, \mathcal{A})$.

The notions of asymptotic sectoriality and quasi-sectoriality are strongly related. In fact, they are equivalent in the following sense. An operator A is in $\text{ASect}(\Omega_{(0,\pi),(\omega,\omega)})$ if an only if $\pm A \in \text{QSect}(i\Sigma_{\pi/2+\omega})$, $0 < \omega < \pi/2$.

In other words, an asymptotically bisectorial operator A may be translated in order to obtain a sectorial operator $A + \nu$ for a suitable $\nu \in \mathbb{C}$. If A has a bounded $H^\infty(\Omega)$ functional calculus and if Σ is an admissible sector for the shifted operator $A + \nu$ with $\Sigma \supset \tilde{\Omega} = \Omega + \nu$, then $A + \nu$ has a bounded $H^\infty(\Sigma)$-functional calculus. In fact, by the homology version of Cauchy's theorem the $H^\infty(\Sigma)$-functional calculus is a restriction of the $H^\infty(\tilde{\Omega})$-functional calculus.

In the following we will transfer the theorem of Kalton and Weis to this setting. The arguments are essentially the same, this time we do not have a singularity at the origin. It remains to use the randomization technique to get the estimates at infinity. This will be done by transferring the problem to the classical sectorial setting by means of suitable translations. Then we will make use of the estimates that have already been proven in this case.

In the following $\mathcal{A}$ denotes the commutator algebra of the operator A.

1.7.17 Theorem. *Let A be asymptotically bisectorial with dense domain. Let $\Omega < \Omega'$ be admissible sets. Assume that A has a bounded $H^\infty(\Omega)$-functional calculus, then A has a bounded $RH^\infty(\Omega', \mathcal{A})$-functional calculus; we write $A \in RH^\infty(\Omega', \mathcal{A})$.*

More precisely, there is a constant $C = C(A) > 0$ such that for all $g \in RH^\infty(\Omega', \mathcal{A})$ we have $\|g(A)\| \le C\mathcal{R}_2(g(\Omega'))$.

Proof. As the argument is quite similar to the one given in the bisectorial case, we will be brief. By the convergence lemma and Kahane's contraction principle it suffices to establish the assertion for functions decaying regularly at infinity. Let α be the direction and ω the opening angle of Ω'. Let Γ be an admissible contour for Ω'. Choose numbers $e^{i(\pi-\eta)}, e^{i(\pi-\mu)}$ in the two connected components of the complement of $\overline{\Sigma_{\alpha,\omega}}$. Then choose a number $\nu > 0$ large enough, such that $\nu + e^{i\eta}A$ and $\nu + e^{i\mu}A$ are invertible, sectorial operators and that $e^{i(\pi-\eta)}\nu, e^{i(\pi-\mu)}\nu$ are not contained in $\overline{\Omega} \cup \Gamma$.

Given $g \in H_d^\infty(\Omega', \mathcal{A})$ with R-bounded range and $r \in (0,1)$ the operator $g(A)$ is given by the integral (the fractional powers are defined using the principal branch of the logarithm)

$$g(A) = \frac{1}{2\pi i} \int_\Gamma g(z)(\nu + e^{i\eta}z)^{-r}(\nu + e^{i\eta}A)^r R(z, A)\, dz.$$

We split Γ into two parts $\Gamma_1 = \Gamma \cap \{z : |z| \le R\}$, $R > \nu$, and Γ_2 such that Γ_2 consists of four half-rays γ_k. We may assume that Γ was chosen such that $\Gamma_2 \subset \Omega' \setminus \overline{\Omega}$.

As the contour Γ_1 has finite length and since the integrand is holomorphic on Γ_1, we find at once that there is a constant $C_1 = C_1(A) > 0$ such that

$$\left\| \frac{1}{2\pi i} \int_{\Gamma_1} g(z)(\nu + e^{i\eta}z)^{-r}(\nu + e^{i\eta}A)^r R(z, A)\, dz \right\| \le C_1 \mathcal{R}_2(g(\Omega')).$$

It remains to estimate the integral over the contour Γ_2 which will be done by estimating the integral along each of the four half-lines γ_k separately using the randomization technique as in the sectorial case.

For two of the four half-lines, say for $k = 1, 2$, the proof of Theorem 1.7.6 yields the following estimates

$$\left\| \int_{\gamma_k} h_\eta(z)(\nu + e^{i\eta}z)^{-r}(\nu + e^{i\eta}A)^r R(z, A)\, dz \right\| \leq C\mathcal{R}_2(h_\eta(\gamma_k))$$

for any function h_η defined on γ_k with R-bounded range and decaying regularly at infinity. Likewise, we obtain for $k = 3, 4$

$$\left\| \int_{\gamma_k} h_\mu(z)(\nu + e^{i\mu}z)^{-r}(\nu + e^{i\mu}A)^r R(z, A)\, dz \right\| \leq C\mathcal{R}_2(h_\mu(\gamma_k))$$

where h_μ is defined on γ_k, has R-bounded range and decays regularly at infinity. We choose $h_\eta(z) = g(z)$ and $h_\mu(z) = g(z)(\nu + e^{i\eta}z)^r(\nu + e^{i\eta}z)^{-r}(\nu + e^{i\eta}A)^r(\nu + e^{i\mu}A)^{-r}$. Note that the operator $(\nu + e^{i\eta}A)^r(\nu + e^{i\mu}A)^{-r}$ is bounded because $A \in H^\infty(\Omega)$. The boundedness of the function $z \mapsto (\nu + e^{i\mu}z)^r(\nu + e^{i\eta}z)^{-r}$ on γ_k, $k = 3, 4$, implies by Kahane's contraction principle the estimate $C\mathcal{R}_2(h_\mu(\gamma_k)) \leq \tilde{C}\mathcal{R}_2(g(\gamma_k))$. Combining these five estimates we obtain the required bound on the norm of $g(A)$ by a multiple of $\mathcal{R}_2(g(\Omega'))$.   □

1.7.18 Remarks.   1. The proof shows that we require only at infinity the assumption $\Omega < \Omega'$. We only made use of the fact that the opening angle of Ω is strictly less then the opening angle of Ω'. More precisely, we only used the assumption $A \in H^\infty(\Omega)$ to deduce that $A \in \mathrm{QH}^\infty(\Sigma)$ for two suitable sectors $\Sigma \supset \nu \pm \overline{\Omega}$ and that the operator $(\nu + e^{i\eta}A)^r(\nu + e^{i\mu}A)^{-r}$ is bounded.

If X is any Banach space and if $\nu + e^{i\eta}A, \nu + e^{i\mu}A \in \mathrm{BIP}$ are invertible, then the operator $(\nu + e^{i\eta}A)^r(\nu + e^{i\mu}A)^{-r}$ is bounded as $\mathcal{D}((\nu + e^{i\eta}A)^r) = [X, \mathcal{D}(\nu + e^{i\eta}A)]_r = [X, \mathcal{D}(A)]_r = \mathcal{D}((\nu + e^{i\mu}A)^r) = \mathcal{R}((\nu + e^{i\mu}A)^{-r})$ (see [115, p.103] or Theorem 1.9.3).

2. Let $\Omega = \Omega_{\alpha,\omega}$, $\alpha = (-\pi/2, \pi/2)$ and $\omega = (\omega_1, \omega_2)$, where $\omega \in [0, \pi/2)^2$. We say that $A \in \mathrm{ASect}_d(\Omega)$ satisfies (P) if

(P) there exist numbers $\nu \geq 0$ and $r \in (0, 1)$ such that $\nu + A$ and $\nu - A$ are invertible, in $\mathrm{Sect}_d(\theta)$ for some $\theta \in (\pi/2, \pi)$, and $(\nu + A)^r(\nu - A)^{-r}$ is bounded.

Hence, if $\pm A \in QH^\infty(\Sigma_\theta)$, then A satisfies (P).

1.7.19 Corollary. *Let $A \in \mathrm{ASect}_d(\Omega_{\alpha,\omega})$. Let $\Omega_{\alpha,\omega}$ be admissible for A with direction $\alpha = (-\pi/2, \pi/2)$ and opening angle $\omega = (\omega_1, \omega_2)$. If $A \in QH^\infty(\Sigma_{\theta_1}) \cap QH^\infty(-\Sigma_{\theta_2})$ with $\omega_1, \omega_2 \in (\max\{\theta_1 - \pi/2, \theta_2 - \pi/2\}, \pi/2)$, then A has a bounded $RH^\infty(\Omega_{\alpha,\omega}, A)$-functional calculus.*

Application: A closed-sum theorem.   We will apply this theorem in order to deduce the closedness of the operator sum $A + B$ in the setting of asymptotically bisectorial operators. The theorem is an analogue of Theorem 1.7.7.

We will describe the setting in the following. Let A and B be two resolvent commuting asymptotically bisectorial operators with dense domain; more precisely let $A \in$ $\mathrm{ASect}(\Omega_{\alpha,\omega^A})$ and $B \in \mathrm{ASect}(\Omega_{\beta,\omega^B})$ where $\alpha = (0,\pi), \beta = (-\pi/2, \pi/2),\ \omega^A = (\omega_A, \omega_A)$, $\omega^B = (\omega_B, \omega_B),\ 0 \leq \omega_A, \omega_B < \pi/2$ for simplicity.

Assume moreover that $\sigma(A) \cap \sigma(-B) = \emptyset$. In [10, Thm. 3.6] it is shown that $A+B$ is closable, moreover, $0 \in \rho(\overline{A+B})$. In fact, the inverse of $\overline{A+B}$ is given by the integral

$$\frac{1}{2\pi i} \int_\Gamma R(z, A) R(z, -B)\, dz$$

for a suitable contour Γ, in analogy to the sectorial case treated in [39].

The key-ingredient in [10] is the construction of a suitable contour Γ. In fact, it is shown (see [10, Lemma 3.1 and following remarks]) that, if A, B satisfy the above assumptions, then there is a contour Γ such that its inside is an admissible set for A with the additional property that the spectrum of $-B$ is contained in the outside. In the Figure 1.5 are sketched the two kind of contours that may occur in this situation. This construction is also the essence of the proof of Lemma 1.7.14.

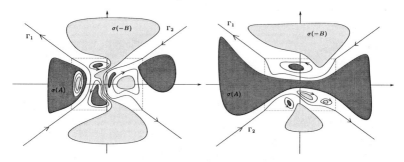

Figure 1.5: In [10] is shown the existence of at least one admissible contour of the type indicated above. The area hatched in dark grey indicates the spectrum of A, the area hatched in light grey the spectrum of $-B$. The contour Γ separates the spectra, the spectrum of A lies in the inside, the spectrum of $-B$ in the outside of the contour. The infinite parts of the contour, Γ_1, Γ_2, are indicated.

We will show that the operator sum $A + B$ with domain $\mathcal{D}(A) \cap \mathcal{D}(B)$ is closed if we require A to have a bounded functional calculus and impose an R-sectoriality condition on B. This will be made precise in the following.

1.7.20 Theorem. *Let A and B be densely defined resolvent commuting operators as described above. Assume that*

1. *A has a bounded $H^\infty(\Omega_A)$-functional calculus,*

2. *$-B \in \mathrm{ARSect}(\Omega_B)$,*

3. $\sigma(A) \cap \sigma(-B) = \emptyset$,

4. $\omega_A + \omega_B < \pi/2$.

Then the operator $A + B$ with domain $\mathcal{D}(A) \cap \mathcal{D}(B)$ is closed and

$$\|Ax\| + \|Bx\| \leq C \|Ax + Bx\|$$

for all $x \in \mathcal{D}(A) \cap \mathcal{D}(B)$. Moreover, $A + B$ is invertible.

Proof. We may choose a slightly bigger quasi-sector $\Omega_A < \Omega'_A$ in such a way that $A \in RH^\infty(\Omega'_A, \mathcal{A})$ by Theorem 1.7.17 and that the two functions $f(z) = B(z + B)^{-1}$ and $g(z) = z(z + B)^{-1}$ lie in $RH^\infty(\Omega'_A, \mathcal{A})$. Hence, from Theorem 1.7.17 we may conclude that the operators $f(A)$ and $g(A)$ are bounded. This implies the closedness of $A + B$ similarly as in the proof of Theorem 1.7.7; observe that we only have to regularize at infinity, hence it suffices to assume dense domains. The fact that $0 \in \rho(A + B)$ is a consequence of the relation $0 \in \rho(\overline{A + B})$ which is proved in [10, Cor. 3.7]. $\square$

Next, we will rephrase this result making use of the concept of quasi-sectoriality.

1.7.21 Corollary. *Let A and B be densely defined resolvent commuting asymptotically bisectorial operators. Assume that*

1. $\pm A \in QH^\infty(i\Sigma_{\theta_A})$;

2. $\pm B \in \mathrm{QRSect}(\Sigma_{\theta_B})$.

If $\theta_A + \theta_B < \frac{3}{2}\pi$, then the operator $A + B$ with domain $\mathcal{D}(A) \cap \mathcal{D}(B)$ is closed and

$$\|Ax\| + \|Bx\| \leq C \|Ax + Bx\|$$

for all $x \in \mathcal{D}(A) \cap \mathcal{D}(B)$. Moreover, $\sigma(A + B) \subset \sigma(A) + \sigma(B)$.

Proof. It suffices to observe that the assumptions imply that $A \in H^\infty(\Omega_A)$ and $B \in \mathrm{ASect}(\Omega_B)$ with Ω_A, Ω_B as in the preceding theorem. Now the claim follows readily from the relations $\theta_A - \pi/2 = \omega_A$ and $\theta_B - \pi/2 = \omega_B$. $\square$

If the underlying Banach space is a Hilbert space, it is well known that the notions of sectoriality and R-sectoriality coincide. Moreover, the angles of sectoriality, R-sectoriality and the angle for the functional calculus agree and are optimal (see [74] and Section 1.10). We formulate this special case in the following corollary.

1.7.22 Corollary. *Let H be a Hilbert space and let A and B be resolvent commuting asymptotically bisectorial operators on H. Assume that $\sigma(A) \cap \sigma(-B) = \emptyset$, $\pm iA \in QH^\infty$, and that $\pm iA \in \mathrm{QSect}(\theta)$, $\pm B \in \mathrm{QSect}(\phi)$ with $\theta + \phi < 3\pi/2$. Then $A + B$ with domain $\mathcal{D}(A) \cap \mathcal{D}(B)$ is closed. Moreover, $0 \in \rho(A + B)$.*

1.8   Real interpolation spaces and Dore's theorem

In this section we consider a single multisectorial operator A. We want to establish a version of Dore's theorem saying that A (or more precisely its part) has a bounded H^∞-functional calculus in real interpolation spaces between X and the domain of A. We assume that the reader is familiar with the definition and basic properties of real interpolation spaces $(X, Y)_{\theta,p}$ constructed by means of the *K-method*. For details and background information we refer to [17], [89]. In our case Y will be the space $\mathcal{D}(A)$ equipped with the graph norm. We begin with a characterization of the interpolation space in terms of the resolvent of A. We denote by $L_*^p(1, \infty)$ the space $L^p((1, \infty), \mu)$ with the measure $\mu = \frac{dt}{t}$.

1.8.1 Proposition. *Let $A \in \mathrm{Sect}(\alpha, \omega)$, $0 < \theta < 1$ and $1 \leq p \leq \infty$. Denote by $(X, \mathcal{D}(A))_{\theta,p}$ the real interpolation space constructed via the K-method, with norm $\|\cdot\|_{\theta,p}$. Let $e^{i\eta}(0, \infty)$ be a ray of minimal growth as in Section 1.5. Then*

$$(X, \mathcal{D}(A))_{\theta,p} = D_A(\theta, p) := \{x \in X : [t^\theta A(e^{i\eta}t - A)^{-1}x] \in L_*^p((1, \infty), X)\} \text{ with}$$

$$\|x\|_{\theta,p} \simeq \|x\| + \left\|t^\theta A(e^{i\eta}t - A)^{-1}x\right\|_{L_*^p}$$

for $x \in D_A(\theta, p)$. If A is invertible, then we even have

$$\|x\|_{\theta,p} \simeq \left\|t^\theta A(e^{i\eta}t - A)^{-1}x\right\|_{L_*^p}.$$

Proof. For a sectorial operator the result is well known and can be found in [63, Chapter 3], [89], or [10]. We apply this to the sectorial operator $-e^{-i\eta}A$ and the result follows. $\square$

This definition is independent of the choice of η, that is, different possible choices of η lead to the same spaces $D_A(\theta, p)$ with equivalent norms, see [10, Section 2].

Clearly, the real interpolation spaces $D_A(\theta, p)$ are invariant under application of the resolvent of A, that is $R(\lambda, A)D_A(\theta, p) \subset D_A(\theta, p)$ with

$$\|R(\lambda, A)x\|_{\theta,p} \leq \|R(\lambda, A)\|_{\mathcal{L}(X)} \|x\|_{\theta,p}$$

for all $\lambda \in \rho(A)$, $0 < \theta < 1$, $1 \leq p \leq \infty$ and $x \in D_A(\theta, p)$. The restriction of A to $D_A(\theta, p)$ is again multisectorial.

1.8.2 Proposition. *Let $A \in \mathrm{Sect}(\alpha, \omega)$, $0 < \theta < 1$ and $1 \leq p \leq \infty$. Denote by $A_{\theta,p}$ the part of A in $D_A(\theta, p)$, denote by $\mathcal{B}$ the commutator algebra of A. Then the following assertions hold.*

1. *$\rho(A) \subset \rho(A_{\theta,p})$ with*

$$R(\lambda, A_{\theta,p}) = R(\lambda, A)|_{D_A(\theta,p)}$$

   for all $\lambda \in \rho(A)$.

2. *$A_{\theta,p} \in \mathrm{Sect}(\alpha, \omega)$ in the Banach space $D_A(\theta, p)$.*

3. *If A is injective or invertible, so is $A_{\theta,p}$.*

4. *If $f \in H_P(\Sigma_{\alpha,\omega'}, \mathcal{B})$ with $\omega < \omega'$ admissible then $f(A_{\theta,p})$ is the part of $f(A)$ in $D_A(\theta, p)$.*

Proof. We refer to [63, III.3.25] for a proof in case A is sectorial and f scalar-valued. The proof in the general setting is essentially the same. □

Now we will state and prove the main result of this section. It generalizes a result of Dore [46] to the setting of multisectorial operators.

1.8.3 Theorem. *Let $A \in \mathrm{Sect}(\alpha, \omega)$ be invertible, $0 < \theta < 1$ and $1 \leq p \leq \infty$. Then for each admissible $\omega' > \omega$ the $H^\infty(\Sigma_{\alpha,\omega'}, \mathcal{B})$-functional calculus for $A_{\theta,p}$ is bounded. In particular, $D_A(\theta, p) \subset \mathcal{D}(f(A))$ for all $f \in H^\infty(\Sigma_{\alpha,\omega'}, \mathcal{B})$.*

Proof. It suffices to prove the result for $p = \infty$. The general case can be deduced from this special case with help of the Reiteration Theorem [89, Section 1.2.3] which states that $D_A(\theta, p)$ is a real interpolation space between $D_A(\theta, \infty)$ and $D_A(\beta, \infty)$, where $\theta < \beta < 1$. Let $p = \infty$. Employing the resolvent equation and the fact that the space $D_A(\theta, p)$ is independent of the choice of η we find that for each contour $\Gamma = \partial\Sigma_{\alpha,\omega''}$ where $\omega'' > \omega$ is admissible there is a constant $c(\omega'')$ such that

$$\sup_{z \in \Gamma} \left\| z^\theta A R(z, A) x \right\| \leq c(\omega'') \left\| x \right\|_{\theta,\infty}$$

for all $x \in D_A(\theta, \infty)$. Here z^θ is defined using an appropriate branch of the logarithm. Then, if $\omega < \omega'' < \omega'$ are admissible, $f \in H^\infty(\Sigma_{\alpha,\omega'}, \mathcal{B})$ and $x \in D_A(\theta, \infty)$, we have

$$\left(\frac{f(z)}{z - e^{i\eta}} \right)(A)x = \frac{1}{2\pi i} \int_\Gamma \frac{f(z)}{z - e^{i\eta}} R(z, A)x \, dz = A^{-1} \frac{1}{2\pi i} \int_\Gamma \frac{f(z)}{z^\theta(z - e^{i\eta})} [z^\theta A R(z, A)x] \, dz.$$

This shows $(f/(z - e^{i\eta}))(A)x \in \mathcal{D}(A) = \mathcal{D}(A - e^{i\eta})$ and thus $x \in \mathcal{D}(f(A))$. Moreover, a direct norm estimate yields

$$
\begin{aligned}
\left\| t^\theta A(e^{i\eta}t - A)^{-1} f(A)x \right\|_X &= \frac{1}{2\pi} \left\| \int_\Gamma \frac{f(z)t^\theta}{z^\theta(e^{i\eta}t - z)} [z^\theta A R(z, A)x] \, dz \right\|_X \\
&\leq c(\omega'') \left\| x \right\|_{\theta,\infty} \frac{1}{2\pi} \int_\Gamma \frac{\left\| f(z) \right\| t^\theta}{|z|^\theta |e^{i\eta}t - z|} \, |dz| \\
&\leq \left\| f \right\|_{H^\infty(\Sigma_{\alpha,\omega'}, \mathcal{B})} c(\omega'') \left\| x \right\|_{\theta,\infty} \frac{1}{2\pi} \int_\Gamma \frac{|dz|}{|z|^\theta |e^{i\eta} - z|}
\end{aligned}
$$

for each $t > 0$. Hence, $f(A)x \in D_A(\theta, \infty)$ with $\left\| f(A)x \right\|_{\theta,\infty} \leq C \left\| f \right\|_{H^\infty(\Sigma_{\alpha,\omega'}, \mathcal{B})} \left\| x \right\|_{\theta,\infty}$ for some constant C independent of x. □

Note, if A is only required to be sectorial, we can always achieve that A^{-1} is bounded by means of a suitable translation. As A is invertible, the resolvent is bounded in a neighborhood about zero. Therefore, we only have to regularize at infinity (see also Remark 1.3.9).

1.9   Bounded imaginary powers

In this section we consider a single injective multisectorial operator A which we will assume to be sectorial. If X is a Hilbert space, Dore's theorem allows to characterize the boundedness of the functional calculus by means of bounded imaginary powers. We recall the definition.

1.9.1 Definition. Let A be a sectorial operator which is densely defined and has dense range. We say that A has *bounded imaginary powers* (BIP) if $A^{is} \in \mathcal{L}(X)$ for all $s \in \mathbb{R}$. We write $A \in$ QBIP if there is $\nu \geq 0$ such that $\nu + A \in$ BIP.

Note, if X is reflexive, we have automatically $\overline{\mathcal{D}(A)} = X$, see [61, Prop. 1.1 h)].

Here A^{is} is defined (as always in the case of sectorial operators) using the principal branch of the logarithm. In the next example we will illustrate how the choice of the branch (or, using other words, the rotation employed to obtain a sectorial operator A) does affect the imaginary powers.

1.9.2 Example. Consider the Hilbert space $X = \mathbb{C}$ and the multisectorial operator A defined by $Ax = -x$. Then the spectrum of the bounded operator A is $\{-1\}$ and the resolvent $R(\lambda, A)$ is just multiplication by $(\lambda + 1)^{-1}$. Clearly, $A \in \text{Sect}(\pi, 0)$. We will consider the operators iA and $-iA$ which are both sectorial. An easy computation gives $(iA)^{is}x = e^{s\pi/2}x$ and $(-iA)^{is}x = e^{-s\pi/2}x$.

At least, both choices lead to bounded imaginary powers. However, this is not true in general as we will see later; in fact, it will turn out that, if the bisectorial operator A has bounded imaginary powers, then so does $-A$ if and only if the associated spectral projection is bounded (see Chapter 2). Observe that there is an example that illustrates the fact that if A has a bounded sectorial functional calculus or is R-sectorial, then this is in general not true for the sectorial operator $-A$ (see Example 1.9.7). However, none of these problems can arise in Hilbert space.

In the following we will denote by $[X, \mathcal{D}(A)]_s$ the complex interpolation space between X and $\mathcal{D}(A)$ (equipped with its graph norm), for details we refer to [17] or [89]. Dore's Theorem implies the following characterization of BIP in Hilbert spaces. A related result is stated in [14].

1.9.3 Theorem. *Let X be a Hilbert space, $A \in \text{Sect}_d(\alpha, \omega)$ be invertible and let $s \in (0, 1)$. The following assertions are equivalent:*

  1. $A \in BIP$;

  2. $\mathcal{D}(A^s) = [X, \mathcal{D}(A)]_s$;

  3. A has a bounded $H^\infty(\Sigma_{\alpha,\omega'})$-functional calculus for one (all) admissible $\omega' > \omega$.

Proof. $1 \Rightarrow 2$ is well known and may be found in [115, p.103]. $3 \Rightarrow 1$ is easy to see. It remains to show $2 \Rightarrow 3$. As X is a Hilbert space, we have $[X, \mathcal{D}(A)]_s = (X, \mathcal{D}(A))_{s,2}$. Therefore, by Dore's theorem the part of A in $\mathcal{D}(A^s)$, which is similar to A by the isomorphism $A^s : \mathcal{D}(A^s) \to X$, has a bounded $H^\infty(\Sigma_{\alpha,\omega'})$-functional calculus. $\square$

1.9.4 Remark. In Section 1.10 we will remove the restriction that A is assumed to be invertible; moreover we will give another characterization of the boundedness of the functional calculus by means of square functions.

In the following let A be an injective and sectorial operator on a Banach space X. If A has BIP, the operators $(A^{is})_{s \in \mathbb{R}}$ form a strongly continuous group. Its *group type*

$$\theta_A = \theta(A^{is}) = \inf\{\theta > 0 : \exists M \geq 1, \ \left\|A^{is}\right\| \leq M e^{\theta|s|} \ \forall s \in \mathbb{R}\}$$

is called the *BIP-type* of A. If A has a bounded H^∞-functional calculus, it is known (see [37]) that the BIP-type θ_A is equal to $\omega_{H^\infty}(A)$ the optimal angle for the calculus. Using this property we obtain the following result.

1.9.5 Lemma. *Let $A \in \mathrm{Sect}_d(0)$ on the Banach space X. Let $\eta = \pi/4$. Then the rotated operators $A_1 = e^{i\eta}A$ and $A_2 = e^{-i\eta}A$ are in $\mathrm{Sect}_d(\eta, 0)$ and $\mathrm{Sect}_d(-\eta, 0)$, respectively. Define on the direct sum $Y = X \oplus X$ the operator $B = A_1 \oplus A_2$. Then it is easy to see that B is in $\mathrm{Sect}_d(\alpha, \omega)$ with $\alpha = (-\eta, \eta)$ and $\omega = (0,0)$. Assume that A has a bounded $H^\infty(\Sigma_{\pi/3})$-functional calculus. Then B has a bounded sectorial $H^\infty(\Sigma_{7\pi/12})$-functional calculus.*

If B has a bounded $H^\infty(\Sigma')$-functional calculus, where $\Sigma' = \Sigma_{7\pi/12} \setminus \overline{\Sigma_\epsilon}$ for some $\epsilon \in [0, \pi/4)$, then $\omega_{H^\infty}(A) \leq \pi/4 - \epsilon$, i.e. A has a bounded $H^\infty(\Sigma_\sigma)$-functional calculus for all $\sigma > \pi/4 - \epsilon$.

Proof. Assume that the $H^\infty(\Sigma')$-functional calculus is bounded for B and hence also for A_1 and A_2. By rotation (using Thm. 1.4.24) and identifying functions defined on a sector with functions on a bisector (extension by zero) we find that A has bounded $H^\infty(\Sigma_i)$ calculi, where $\Sigma_1 = \{re^{is} : r > 0, \ s \in (-\pi/4 + \epsilon, \pi/3)\}$ and $\Sigma_2 = \{re^{is} : r > 0, \ s \in (-\pi/3, \pi/4 - \epsilon)\}$. We use this fact to find a bound for the BIP-type of A: If $s > 0$ we have $\|A^{is}\| \leq C \|z^{is}\|_{H^\infty(\Sigma_1)} = C e^{-s(-\pi/4+\epsilon)} = C e^{|s|(\pi/4-\epsilon)}$. Similarly, for $s < 0$ we obtain $\|A^{is}\| \leq C \|z^{is}\|_{H^\infty(\Sigma_2)} \leq C e^{|s|(\pi/4-\epsilon)}$. Therefore, the $\theta_A \leq \pi/4 - \epsilon$. $\qquad\square$

The actual values of the angles are not important, the key observation is that a bounded bisectorial calculus for the operator B allows to reduce the angle for the sectorial functional calculus for A; provided that the angle for the calculus is greater than η. In the above example we have $\eta = \pi/4 < \pi/3 \leq \omega_{H^\infty}(A)$.

We recall an example due to Kalton [71] of a sectorial operator whose sectoriality angle is strictly less than its H^∞-angle.

1.9.6 Theorem (Kalton). *Let $\theta \in (0, \pi)$, then there exists a Banach space X (depending on θ) and an operator A defined on X with the following properties:*

1. *$A \in \mathrm{Sect}_d(0)$;*

2. *$\omega_{H^\infty}(A) = \theta$.*

There is even an example, where the Banach space X has property (Δ). In fact, in [76] Kalton and Weis construct such an operator on a subspace of L_p.

1.9.7 Example. Let $\theta \in (\pi/3, \pi)$ and let A and X be as in Kalton's example (Thm. 1.9.6); assume that X has property (Δ). Then the operator B defined in Lemma 1.9.5 is bisectorial with dense range and domain, has a bounded sectorial H^∞-functional calculus, but B has not a bounded H^∞-functional calculus on a bisector around the real line. Indeed, if it had such a bounded bisectorial functional calculus, we would have $\omega_{H^\infty}(A) \leq \pi/4 < \theta$, by Lemma 1.9.5, a contradiction.

We will see in the following section that this implies that $-B$ has neither a bounded sectorial functional calculus nor is it (almost) R-sectorial.

1.10 Square functions

In this section we will consider a single multisectorial injective operator $A \in \mathrm{Sect}_d(\alpha, \omega)$ with $\omega \in [0, \pi)^N$ admissible for $\alpha \in (-\pi, \pi]^N$. We write $\Sigma = \Sigma_{\alpha,\omega}$ and $\Sigma' = \Sigma_{\alpha,\omega'}$ for $\omega' > \omega$. We will characterize the boundedness of the H^∞-functional calculus by means of square functions. Originally, square functions come from harmonic analysis and were introduced in operator theory by McIntosh and Yagi (see [90] and [125]). They studied in case of a sectorial operator on a Hilbert space the expression

$$\|x\|_\phi := \int_0^\infty \|\phi(tA)x\|^2 \, \frac{dt}{t} \tag{1.5}$$

for $\phi \in H_0^\infty(\Sigma')$. The main point is that different choices of ϕ give rise to equivalent norms.

In each of the N connected components of $\mathbb{C} \setminus \overline{\Sigma'}$ let us choose a point $e^{i\eta_k}$, where $\eta_k \in (-\pi, \pi]$. We will denote the sectorial operator $e^{i(\pi-\eta_k)}A$ by A_k and its sectoriality angle by $\omega(A_k) \in [0, \pi)$.

1.10.1 Hilbert spaces

At first we will consider the more elementary case of A being defined on a Hilbert space H. We will not transport all the theory of square functions to the multisectorial case, but restrict ourselves to a special case of concrete square functions associated to the functions of the form $\phi_{\nu,\theta} : z \mapsto (e^{i\theta}z)^{1/2}(e^{i\nu} - z)^{-1}$.

A simple change of variables gives $\int_0^\infty \|\phi_{\nu,\pi-\eta_k}(tA)x\|^2 \, \frac{dt}{t} = \int_0^\infty \left\| A_k^{1/2} R(se^{i\nu}, A)x \right\|^2 ds$.

First, we will see that all the square functions that we will consider in the following are equivalent. In fact, given $\mu, \nu \in (-\pi, \pi]$ such that $e^{i\mu}, e^{i\nu} \notin \overline{\Sigma}$ we have

$$\int_0^\infty \left\| A_k^{1/2} R(te^{i\mu}, A)x \right\|^2 dt \sim \int_0^\infty \left\| A_k^{1/2} R(te^{i\nu}, A)x \right\|^2 dt.$$

Indeed, the resolvent equation gives

$$A_k^{1/2} R(te^{i\mu}, A) = [I + t(e^{i\mu} - e^{i\nu})R(te^{i\mu}, A)]A_k^{1/2} R(te^{i\nu}, A).$$

As the term in the squared brackets is bounded on $(0, \infty)$, we obtain an inequality; by symmetry in ν and μ we obtain equivalence. The same argument is valid if we replace A by its adjoint A^*.

1.10.1 Theorem. *Let $A \in \mathrm{Sect}_d(\alpha, \omega)$. The following statements are equivalent:*

(H1) A has a bounded $H^\infty(\Sigma_{\alpha, \omega'})$-functional calculus for one (all) admissible $\omega' > \omega_A$;

(H2) A_k has bounded imaginary powers for all k and for one (all) $\nu_k \in (\omega(A_k), \pi]$ there is a constant $C > 0$ with
$$\left\| A_k^{is} \right\| \leq C e^{\nu_k |s|};$$

(H3) A_1 has bounded imaginary powers and satisfies the estimate given in (H2);

(H4) For one (all) $\mu \in (-\pi, \pi]$ such that $e^{i\mu} \notin \overline{\Sigma}$ there is a constant $C > 0$ such that
$$\left\| A_1^{1/2} R(\cdot e^{i\mu}, A)x \right\|_{L_2((0,\infty),H)} \leq C \|x\|,$$
$$\left\| (A_1^*)^{1/2} R(\cdot e^{i\mu}, A^*)x \right\|_{L_2((0,\infty),H)} \leq C \|x\|;$$

(H5) For one (all) $\mu \in (-\pi, \pi]$ such that $e^{i\mu} \notin \overline{\Sigma}$ there is a constant $C > 0$ such that
$$C^{-1} \|x\| \leq \left\| A_1^{1/2} R(\cdot e^{i\mu}, A)x \right\|_{L_2((0,\infty),H)} \leq C \|x\|.$$

Proof. We first note that conditions (H4) and (H5) hold for all admissible μ if they hold for one such ν as was shown above.

$(H1) \Rightarrow (H2)$: This follows immediately as $A_k = e^{i(\pi - \eta_k)}A$ has a bounded $H^\infty(\Sigma_{\nu_k})$-functional calculus (compare Thm. 1.4.24) and from the estimate $|z^{is}| = e^{-i(\arg z)s} \leq e^{\nu_k |s|}$ for $z \notin \Sigma_{\nu_k}$.

$(H2) \Rightarrow (H3)$: This is trivial.

$(H3) \Rightarrow (H4)$: A proof for the case of a sectorial operator (i.e. $A = A_1$) can be found in [81, Thm. 11.9]. Now it suffices to observe that
$$R(te^{i\mu}, e^{i(\pi - \eta_1)}A) = e^{-i(\pi - \eta_1)} R(te^{i(\mu - \pi + \eta_1)}, A)$$

and that $e^{i\mu}(0, \infty)$ is a ray of minimal growth for $A_1 = e^{i(\pi - \eta_1)}A$ if and only if $e^{i(\mu - \pi + \eta_1)}(0, \infty)$ is a ray of minimal growth for A. Hence
$$\left\| A_1^{1/2} R(\cdot e^{i(\mu - \pi + \eta_1)}, A)x \right\|_{L_2((0,\infty),H)} = \left\| A_1^{1/2} R(\cdot e^{i\mu}, A_1)x \right\|_{L_2((0,\infty),H)} \leq C \|x\|;$$

the same arguments apply to A^*.

$(H4) \Rightarrow (H5)$: Is deduced from the sectorial case as in the implication $(H3) \Rightarrow H(4)$; alternatively one can proceed as in [81].

$(H5) \Rightarrow (H1)$: We follow the argument in [81]. Let $\omega' > \omega$ be admissible for A, $\Gamma \subset \Sigma' \setminus \overline{\Sigma}$ be an admissible contour and $f \in H_0^\infty(\Sigma')$. Decompose Γ into a sum of

N contours $\gamma_k = \partial\Sigma_{\alpha_k,\bar{\omega}_k}$ with $\omega_k < \tilde{\omega}_k < \omega'_k$. We will estimate the norm of $f(A)$ by "pushing" $f(A)$ through the square function. Note that assumption (H5) together with the independence on μ gives for $1 \le k \le N$

$$\left\| A_1^{1/2} R(\cdot, A)x \right\|_{L_2(\gamma_k, H)} \sim \|x\|_H. \tag{1.6}$$

Writing

$$f_k(A) = \frac{1}{2\pi i} \int_{\gamma_k} f(z) R(z, A)\, dz$$

we calculate for $w \in \gamma_k$, using the resolvent equation,

$$A_1^{1/2} R(w, A) f_k(A) = (2\pi i)^{-1} \int_{\gamma_k} f(z) A_1^{1/2} R(w, A) R(z, A)\, dz$$

$$= \left((2\pi i)^{-1}\text{PV-}\int_{\gamma_k} \frac{f(z)}{z - w}\, dz \right) A_1^{1/2} R(w, A) - (2\pi i)^{-1}\text{PV-}\int_{\gamma_k} \frac{f(z) A_1^{1/2} R(z, A)}{z - w}\, dz$$

$$= \frac{f(w)}{2} A_1^{1/2} R(w, A) - K_k[f(\cdot) A_1^{1/2} R(\cdot, A)](w) \tag{1.7}$$

by Cauchy's theorem and using the notation

$$K_k G(w) = (2\pi i)^{-1}\text{PV-}\int_{\gamma_k} \frac{G(z)}{z - w}\, dz \qquad w \in \gamma_k.$$

K_k is a variant of the Hilbert transform and it is bounded on $L_2(\gamma_k)$. By [81, 11.12 b)], we have for $g \in L_2(\gamma_k, H)$

$$\int_{\gamma_k} \|K_k g(w)\|^2\, d|w| \le M_k^2 \int_{\gamma_k} \|g(w)\|^2\, d|w|.$$

Then, for $x \in \mathcal{D}(A)$ it follows from (1.6) and (1.7) that

$$\|f_k(A)x\|_H \le C \left\| A_1^{1/2} R(\cdot, A)[f_k(A)x] \right\|_{L_2(\gamma_k, H)}$$

$$\le C/2 \left\| f(\cdot) A_1^{1/2} R(\cdot, A)x \right\|_{L_2(\gamma_k, H)} + C M_k \left\| f(\cdot) A_1^{1/2} R(\cdot, A)x \right\|_{L_2(\gamma_k, H)}$$

$$\le C(1/2 + M_k) \|f\|_{H^\infty(\Sigma')} \left\| A_1^{1/2} R(\cdot, A)x \right\|_{L_2(\gamma_k, H)} \le C_k \|f\|_{H^\infty(\Sigma')} \|x\|_H$$

where in the last step we used again (1.6). Therefore, by the triangle inequality

$$\|f(A)x\|_H \le \tilde{C} \|f\|_{H^\infty(\Sigma')} \|x\|_H$$

which proves the claim. $\square$

1.10.2 Remark. For a sectorial operator A defined on a general Banach space having BIP is not equivalent to possessing a bounded H^∞-functional calculus. In fact, for $2 \neq p \in (1, \infty)$ Lancien constructed an operator on L_p having BIP but no bounded functional calculus (see [84]). If A has a bounded functional calculus, then the set $\{A^{is} : s \in [-1, 1]\}$ is clearly bounded; if, in addition, the Banach space X has property (α), it is even R-bounded (by Thm. 1.7.10). The converse holds, if X has finite cotype (e.g. if X is uniformly convex); in this case the R-boundedness of the set $\{A^{is} : s \in [-1, 1]\}$ is equivalent to the boundedness of the functional calculus (see [75]).

1.10.2 Banach spaces

In a general Banach space the definition of a square function given by (1.5) does not lead to the desired results. The spaces obtained by means of these norms are called McIntosh-Yagi spaces in [61]. They are closely related to the real interpolation spaces between $\mathcal{R}(A)$ and $\mathcal{D}(A)$ (with homogeneous norm, see [61, 63]).

Thus, we are seeking to replace it by something more suitable. If X is a Hilbert space, an easy calculation, using the fact that $r_k(\cdot)\phi(2^k tA)x$ is an orthogonal sequence in the Hilbert space $L_2(X) := L_2([0, 1], X)$, gives

$$\|x\|_\phi^2 = \int_0^2 \left(\int_0^1 \left\| \sum_{k \in \mathbb{Z}} r_k(u)\phi(2^k tA)x \right\|^2 du \right) \frac{dt}{t}$$

for all $\phi \in H_0^\infty(\Sigma')$. The map $t \mapsto \phi(tA)$ is continuous and hence the integral makes sense. For this and other properties of this map we refer to [90], [61, 4.4], [75] and [81].

This motivates the following definition [72].

1.10.3 Definition. Let $A \in \text{Sect}_d(\alpha, \omega)$ and $\phi \in H_0^\infty(\Sigma_{\alpha,\omega'})$, $\omega < \omega'$. We define

$$\|x\|_\phi = \sup_{t>0} \left\| \sum_{k \in \mathbb{Z}} r_k \phi(t2^k A)x \right\|_{L_p(X)} \tag{1.8}$$

for $x \in X$. In particular, for $\phi = \phi_{\nu,\theta}$ we have

$$\|x\|_{\phi_{\nu,\theta}} = \sup_{t>0} \left\| \sum_{k \in \mathbb{Z}} r_k (e^{i\theta}A)^{1/2} R(t2^k e^{i\nu}, A)x \right\|_{L_p(X)} .$$

We will write $\|\cdot\|_{\phi,A}$ if we want to distinguish the operator A with which the square function is associated. Different choices of $p \in [1, \infty)$ give rise to equivalent norms.

The argument linking square function estimates with the H^∞-functional calculus will be based on the following result which states the equivalence of those norms. In the sectorial case it is clear that we have to require $\phi \neq 0$. Similarly, in the general case, we demand that ϕ vanishes on none of the connected components of the multisector. To this end denote by $\mathcal{N}$ the set $\{f \in H^\infty(\Sigma_{\alpha,\omega'}) : f$ vanishes on a non-empty open subset$\}$; in case of a single sector the set $\mathcal{N}$ reduces to $\{0\}$. The following result is an extension of [81, Prop. 12.15] to the setting of multisectorial operators.

1.10.4 Proposition. *Let $A \in \operatorname{Sect_d}(\alpha, \omega)$ on the Banach space X. For every pair of functions $\phi, \psi \in H_0^\infty(\Sigma_{\alpha,\omega'}) \setminus \mathcal{N}$, $\omega < \omega'$, there is a constant $C > 0$ such that for all $x \in X$*

$$C^{-1} \|x\|_\phi \leq \|x\|_\psi \leq C \|x\|_\phi.$$

Proof. The argument in the sectorial case carries over with only minor changes. Therefore, we outline only the points that need to be changed; for a proof in the sectorial setting we refer to [81, Prop. 12.15] for the Banach space case or [90] for the Hilbert space case.

Clearly, it suffices to establish only the inequality $\|x\|_\psi \leq C \|x\|_\phi$ as we may interchange the roles of ψ and ϕ. By our assumption and the identity theorem for holomorphic functions we may find an auxiliary function $g \in H_0^\infty(\Sigma_{\alpha,\omega'})$ such that

$$\int_0^\infty g(\lambda t)\phi(\lambda t)\frac{dt}{t} = 1$$

for all $\lambda \in \Sigma_{\alpha,\omega'}$; note that the measure $\frac{dt}{t}$ is invariant with respect to scaling. This implies by [81, Lemma 9.13] that for all $x \in \mathcal{D}(A) \cap \mathcal{R}(A)$

$$\int_0^\infty g(tA)\phi(tA)x\frac{dt}{t} = x. \tag{1.9}$$

Having established this relation the proof is completed as in [81, Prop. 12.15]. □

Using this equivalence we will obtain the desired characterization of the boundedness of the functional calculus in terms of square functions. However, we first introduce the following notion, which weakens the concept of R-sectoriality. It was introduced by Kalton, Kunstmann and Weis in [72].

1.10.5 Definition. Let $A \in \operatorname{Sect}(\alpha, \omega)$ be a sectorial operator on the Banach space X. We say that A is *almost R-sectorial* if there is an admissible $\omega < \tilde{\omega}$ such that the set $\{\lambda AR(\lambda, A)^2 : \lambda \in \mathbb{C} \setminus \overline{\Sigma_{\alpha,\tilde{\omega}}}\}$ is R-bounded. We will write $A \in \operatorname{aRSect}(\alpha, \tilde{\omega})$ or $A \in \operatorname{aRSect}(\Sigma_{\alpha,\tilde{\omega}})$ in this case. The infimum of all such $\tilde{\omega}$ is called the angle of almost R-sectoriality and denoted by $\omega_r(A)$. We write $A \in \operatorname{aRSect_d}(\alpha, \tilde{\omega})$ to indicate that A is moreover densely defined with dense range.

The factorization $\lambda AR(\lambda, A)^2 = [\lambda R(\lambda, A)][AR(\lambda, A)]$ shows that the notion of almost R-sectoriality is indeed weaker than that of R-sectoriality. In general, it is strictly weaker; there are even examples on L_p, $p \neq 2$ (see [72]). This notion is important as, if A has a bounded functional calculus, then it is almost R-sectorial, whereas it is in general not R-sectorial unless the Banach space satisfies property (Δ). If A is sectorial, we have the following reformulation of almost R-sectoriality.

1.10.6 Lemma. *Let $A \in \operatorname{aRSect}(\theta)$, $\theta \in [0, \pi)$ and $s \in (0, 1)$, then for all $\theta' > \theta$*

$$\mathcal{R}\{\lambda^s A^{1-s}(\lambda + A)^{-1} : \lambda \in \Sigma_{\pi-\theta'}\} < \infty.$$

Proof. Fix $\phi \in (\pi - \theta', \pi - \theta)$ and consider the functions $\psi(z) = z^{1-s}(1 + e^{\pm i\phi}z)^{-1} \in H_0^\infty(\Sigma_{\pi-\theta'})$. Then, for $t > 0$, we have $\psi(t^{-1}A) = e^{\mp i\phi}t^s A^{1-s}(te^{\mp i\phi} + A)^{-1}$. We deduce from the following proposition the R-boundedness of the set $\{\lambda^s A^{1-s}(\lambda+A)^{-1} : 0 \neq \lambda \in \partial\Sigma_\phi\}$. An application of Lemma 1.6.6 now yields the claim. $\qquad\square$

The following result is taken from [81, Lemma 12.5].

1.10.7 Proposition. *Let* $A \in \mathrm{aRSect}(\theta)$ *and* $\sigma > \theta$. *For every* $\psi \in H_0^\infty(\Sigma_\sigma)$ *there is a* $\varphi \in H_0^\infty(\Sigma_\sigma)$ *and a constant* γ *such that for* $\sigma > \nu > \theta$ *and* $t > 0$

$$\psi(tA) = \frac{1}{2\pi i}\int_{\partial\Sigma_\nu} \frac{\varphi(\lambda)}{\lambda}[tA\lambda R(\lambda, tA)^2]d\lambda + \gamma tA(1 + tA)^{-2}.$$

In particular, the set $\{\psi(tA) : t > 0\}$ *is R-bounded.*

Proof. Let $\bar\psi$ be an antiderivative of $\lambda^{-1}\psi(\lambda)$ which vanishes at 0. Define $\varphi(\lambda) = \bar\psi(\lambda) - \gamma\lambda(1 + \lambda)^{-1}$, where $\gamma = \int_0^\infty t^{-1}\psi(t)dt$. Then

$$\varphi'(\lambda) = \lambda^{-1}\psi(\lambda) - \gamma(1 + \lambda)^{-2}. \tag{1.10}$$

The function φ belongs to $H_0^\infty(\Sigma_\sigma)$. Hence, for $\sigma > \nu > \theta$

$$\varphi(tA) = \frac{1}{2\pi i}\int_{\partial\Sigma_\nu} \varphi(t\lambda)R(\lambda, A)d\lambda = \frac{1}{2\pi i}\int_{\partial\Sigma_\nu} \varphi(\lambda)R(\lambda, tA)d\lambda.$$

Therefore

$$\frac{d}{dt}\varphi(tA) = \frac{1}{2\pi i}\int_{\partial\Sigma_\nu} \lambda\varphi'(t\lambda)R(\lambda, A)d\lambda = A\varphi'(tA)$$

moreover

$$\frac{d}{dt}\varphi(tA) = \frac{1}{2\pi i}\int_{\partial\Sigma_\nu} \varphi(\lambda)[AR(\lambda, tA)^2]d\lambda.$$

Applying the functional calculus to the identity (1.10) gives

$$\psi(tA) - \gamma tA(1 + tA)^{-2} = (tA)\varphi'(tA) = \frac{1}{2\pi i}\int_{\partial\Sigma_\nu} \varphi(\lambda)[tAR(\lambda, tA)^2]d\lambda.$$

Now, the R-boundedness of the set $\{\psi(tA) : t > 0\}$ follows by Corollary 1.6.5. $\qquad\square$

1.10.8 Remark. Observe, that, if the operator A is multisectorial, we have to substitute the function $\gamma(1+\lambda)^{-2}$ by $(1+\lambda)^{-2}(\sum_{k=1}^N \gamma_k p_k)$, where p_k denotes the indicator function of the sector $\Sigma_{\alpha_k, \omega_k'}$. Therefore, the question is related to the boundedness of the spectral projections $P_k = p_k(A)$ in this case.

The following proposition stated in [72] relates the notions of bounded imaginary powers and almost R-sectoriality.

1.10.9 Proposition. *Let* A *be a sectorial operator with bounded imaginary powers and BIP-type* $\theta_A < \pi$. *Then* A *is almost R-sectorial with* $\omega_r(A) \leq \theta_A$.

Proof. See [72, Proposition 3.2].   □

Note, if A is sectorial and admits a bounded functional calculus, then the assumptions of the preceding proposition are satisfied. Indeed, in this case the BIP-type equals the angle of the calculus which is less than π. Hence, by suitable rotations, we obtain at once the following corollary.

1.10.10 Corollary. *Let $A \in \mathrm{Sect}_\mathrm{d}(\alpha, \omega)$. Assume that A has a bounded $H^\infty(\Sigma_{\alpha,\omega'})$-functional calculus. Then A is almost R-sectorial with angle $\omega_r(A) \leq \omega'$. In particular, $\omega_r(A) \leq \omega_{H^\infty}(A)$.*

In general, the operator A is not R-sectorial; however, if the underlying Banach space satisfies the geometric property (Δ), then we may replace the notion of almost R-sectoriality by R-sectoriality in the foregoing statement (see [74, Thm. 5.3]).

Now we will combine the notion of almost R-sectoriality with the tool of square functions. This will tell us precisely, how, given an operator with a bounded functional calculus, we may reduce its domain of holomorphy such that the functional calculus still remains bounded. This is the content of the following theorem.

In the theorem, we require that the set $\{z^{1/2}A^{1/2}R(z,A) : z \in \mathbb{C} \setminus \overline{\Sigma_{\alpha,\omega}}\}$ to be R-bounded, where $z^{1/2}$ and $A^{1/2}$ are defined by a suitable branch of the logarithm. The next lemma states that this holds if the operator A is R-sectorial. Clearly, it is necessary for A to be almost R-sectorial. By Proposition 2.5.3 this is also sufficient, if we require in addition the boundedness of all associated spectral projections. Observe that this additional assumption is void, if A is a (possibly rotated) sectorial operator (compare with Proposition 1.10.7).

1.10.11 Lemma. *Let $A \in \mathrm{RSect}(\alpha, \omega)$ and let $\tilde{\omega} > \omega$. Then the set $\{z^{1/2}A^{1/2}R(z,A) : z \in \mathbb{C} \setminus \overline{\Sigma_{\alpha,\tilde{\omega}}}\}$ is R-bounded.*

Proof. Let $z \in \mathbb{C} \setminus \overline{\Sigma_{\alpha,\tilde{\omega}}}$. The operator $z^{1/2}A^{1/2}R(z,A)$ is given by the contour integral

$$\frac{1}{2\pi i} \int_\Gamma \frac{z^{1/2}\lambda^{1/2}}{z - \lambda} R(\lambda, A)\, d\lambda = \frac{1}{2\pi i} \int_\Gamma h(z, \lambda)[\lambda R(\lambda, A)]\, \frac{d\lambda}{\lambda}$$

where $\Gamma \subset \Sigma_{\alpha,\tilde{\omega}} \setminus \overline{\Sigma_{\alpha,\omega}}$ is an admissible contour and $h(z, \lambda) = \frac{z^{1/2}\lambda^{1/2}}{z - \lambda}$ is homogeneous of degree 0. Therefore, it suffices to consider only those z that lie in the compact set $\{w \in \mathbb{C} : |w| = 1\} \setminus \Sigma_{\alpha,\tilde{\omega}}$. Hence, the claim is a consequence of the uniform integrability of $|h(z, \lambda)\lambda^{-1}|$ and Corollary 1.6.5.   □

The following two theorems are extensions of results from [72, 81] to the setting of multisectorial operators.

1.10.12 Theorem. *Let $A \in \mathrm{Sect}_\mathrm{d}(\alpha, \omega)$.*

    1. *If A has a bounded $H^\infty(\Sigma_{\alpha,\omega'})$-functional calculus for $\omega' > \omega$ and if $\phi \in H_0^\infty(\Sigma_{\alpha,\sigma}) \setminus \mathcal{N}$, $\sigma > \omega'$, then there is a constant C such that*

$$C^{-1}\|x\| \leq \|x\|_\phi \leq C\|x\| \qquad \text{for all } x \in X. \tag{1.11}$$

2. *Conversely, if the set*

$$\{z^{1/2}A^{1/2}R(z,A) : z \in \mathbb{C} \setminus \overline{\Sigma_{\alpha,\tilde{\omega}}}\} \tag{1.12}$$

is R-bounded and if (1.11) holds for some $\phi \in H_0^\infty(\Sigma_{\alpha,\tilde{\omega}'}) \setminus \mathcal{N}$ with $\tilde{\omega}' > \tilde{\omega}$ then A has a bounded $H^\infty(\Sigma_{\alpha,\tilde{\sigma}})$-functional calculus for each $\tilde{\sigma} > \tilde{\omega}'$.

In particular, if A has a bounded H^∞-functional calculus, then $\omega_{H^\infty}(A) \le \tilde{\omega}$.

Proof. The argument is as in the sectorial case. Hence, we will be brief and refer to [81, Thm. 12.17] for further details in the sectorial case. Let A have a bounded $H^\infty(\Sigma_{\alpha,\omega'})$-functional calculus. Given $\phi \in H_0^\infty(\Sigma_{\alpha,\sigma}) \setminus \mathcal{N}$ we find again an auxiliary function g such that the relation (1.9) holds. From this we obtain for $x \in X$, $x^* \in X^*$

$$|\langle x, x^* \rangle| \le \int_0^\infty |\langle \phi(tA)x, g(tA)^*x^* \rangle| \frac{dt}{t}$$

which may be estimated as in the sectorial case (using an adaptation of [81, Thm. 12.2 and Lemma 12.6 b)]) by

$$\le 2 \|x\|_\phi \|x^*\|_g \le C \|x\|_\phi \|x^*\| ;$$

and hence $\|x\| \le C \|x\|_\phi$. The reverse inequality is shown similarly.

Let us now consider the converse statement. First note that there is a constant $C > 0$ such that

$$\mathcal{R}(\{f(A)\phi(tA) : t > 0\}) \le C \|f\|_{H^\infty(\Sigma_{\alpha,\tilde{\sigma}})} \tag{1.13}$$

for all $f \in H_0^\infty(\Sigma_{\alpha,\tilde{\sigma}})$. Indeed, if $\Gamma \subset \Sigma_{\alpha,\tilde{\sigma}} \setminus \overline{\Sigma_{\alpha,\tilde{\omega}'}}$ is an admissible contour, we have by Theorem 1.7.3

$$f(A)\phi(tA) = \frac{1}{2\pi i} \int_\Gamma z^{-1}\phi(tz)[f(z)z^{1/2}A^{1/2}R(z,A)]\,dz$$

where $z^{1/2}$ and $A^{1/2}$ are defined by means of a suitable branch of the logarithm. Given our assumption, Kahane's contraction principle implies that the set $\{f(z)z^{1/2}A^{1/2}R(z,A) : z \in \Gamma\}$ is R-bounded; now the claim follows from Corollary 1.6.5. Denote by Σ' the set $\Sigma_{\alpha,\tilde{\sigma}}$. Using the equivalence of different square functions stated in Proposition 1.10.4 we have by (1.13)

$$\|f(A)x\|_X \le C \|f(A)x\|_\phi \le C_1 \|f(A)x\|_{\phi^2} = C_1 \sup_{t>0} \left\| \sum_k r_k[f(A)\phi(t2^kA)]\phi(t2^kA)x \right\|_{L_p(X)}$$

$$\le C_2 \|f\|_{H^\infty(\Sigma')} \sup_{t>0} \left\| \sum_k r_k\phi(t2^kA)x \right\|_{L_p(X)} = C_2 \|f\|_{H^\infty(\Sigma')} \|x\|_\phi \le C_3 \|f\|_{H^\infty(\Sigma')} \|x\|_X$$

which proves the converse. $\qquad\square$

We will show in Section 2.5 (see Proposition 2.5.5) that, given the equivalence (1.11) of the square function norm, the almost R-sectoriality is in effect equivalent to the R-boundedness assumption (1.12) required in Theorem 1.10.12. Although the proof of this fact is quite easy, we postpone it, as it makes use of some properties of spectral projections that we will discuss in detail in the next chapter. We note some of the consequences already here for purposes of reference.

1.10.13 Theorem. *Let $A \in \mathrm{Sect_d}(\alpha, \omega) \cap \mathrm{aRSect}(\alpha, \tilde{\omega})$, $\omega \leq \tilde{\omega}$ admissible. Then A has a bounded functional calculus if and only if for one (all) square functions we have the equivalence (1.11). In this case we have $\omega_{H^\infty}(A) = \omega_r(A)$.*

Proof. A proof is given in Section 2.5; see Theorem 2.5.6. $\qquad\qquad\qquad\qquad$ $\square$

1.10.14 Corollary. *Let $A \in \mathrm{Sect_d}(\alpha, \omega)$ and assume that A admits a bounded $H^\infty(\Sigma_{\alpha,\omega'})$-functional calculus. Then $\omega_{H^\infty}(A) = \omega_r(A)$.*

1.10.15 Remarks.   1. This has the following remarkable consequence; regarding the multisectorial operator A as a sectorial operator (after a suitable rotation) we find that A has a bounded (multisectorial) H^∞-functional calculus if and only if $e^{i(\pi - \eta_k)}A$ has a bounded sectorial H^∞-functional calculus for all $k \in \{1, \ldots, N\}$.

If we know A to be almost R-sectorial, this statement is equivalent to the boundedness of the sectorial functional calculus for a *single* $k \in \{1, \ldots, N\}$.

2. The essential ingredient is, of course, the R-boundedness assumption. Example 1.9.7 exhibits that the second equivalence fails in general. In particular, we find that the bisectorial operator B constructed there is not almost R-bisectorial. However, the operator B is R-sectorial, since it has a bounded functional calculus and X has property (Δ) (see [74, Thm. 5.3]).

3. The coincidence of the angles $\omega_{H^\infty}(A) = \omega_r(A)$ presents an alternative approach to Example 1.9.7 not making use of Lemma 1.9.5 (where we exploited the equality $\omega_{H^\infty}(A) = \theta_A$). Indeed, it suffices to observe that the bisectorial operator B cannot be almost R-bisectorial. If it were so, we would obtain immediately by Corollary 2.5.1 that we could improve the angle of almost R-sectoriality of A to some angle $\tilde{\omega} < \omega_r(A)$, which is a contradiction.

1.10.16 Corollary. *If $A \in H^\infty(\Sigma) \cap H^\infty(\Sigma')$, then $A \in H^\infty(\tilde{\Sigma})$ for all $\tilde{\Sigma} > \Sigma \cap \Sigma'$.*

In the following we will give the full details for a 1-bisectorial operator. Recall that an operator A is called 1-bisectorial if both A and $-A$ are sectorial.

1.10.17 Corollary. *Let A be a densely defined 1-bisectorial operator with dense range defined on a Banach space X. Then A has a bounded bisectorial H^∞-functional calculus if and only if $\pm A$ has a bounded sectorial functional calculus. More precisely, let $\theta > 0$. To $\theta < \theta' < \pi$ we associate the bisector $\Sigma^{\theta'} := \Sigma_{\theta'} \cap (-\Sigma_{\theta'})$.*
    Then $A \in H^\infty(\Sigma^{\theta'})$ for all $\theta' > \theta$ if and only if $\pm A \in H^\infty(\Sigma_{\theta'})$ for all $\theta' > \theta$.

Chapter 2

Spectral decomposition

In this chapter we will use the tools provided by the functional calculus in order to study the existence of spectral decompositions and their associated spectral projections of multisectorial operators.

2.1 Introduction

Consider, for simplicity, a bisectorial operator A with its spectrum in a double cone around the imaginary axis, i.e. A is 1-bisectorial. The spectral projections P_1, P_2 corresponding to the upper and lower cone in general cannot be defined by the Dunford calculus and, indeed, they may be unbounded. E.g. for $A = \frac{d}{dx}$ on $L_p(\mathbb{R})$ the projection P_1 is only bounded for $1 < p < \infty$ and projects $L_p(\mathbb{R})$ onto the Hardy space H_p. However, the boundedness of these spectral projections is closely connected to some important inequalities in analysis. We study these questions in the framework of the H^∞-calculus developed in the first chapter. This allows us to give more direct proofs for some known results, improve on them and also to discover new connections. The functional calculus constructed in the first chapter provides a convenient and unifying framework for all results in this chapter.

E.g. it is a result of Dore and Venni [49] that, for $0 \in \rho(A)$, the boundedness of these spectral projections is equivalent to the "symmetry" of the operators A and $-A$ in the sense that for $w \in \mathbb{C} \setminus \mathbb{Z}$ we have

$$\|A^w x\| \approx \|(-A)^w x\| \qquad \text{for } x \in \mathcal{D}(A^w) = \mathcal{D}((-A)^w).$$

We will give a different proof using the H^∞-calculus that also works for $0 \notin \rho(A)$.

More importantly, the boundedness of the spectral projection is equivalent to the estimate (see Section 3)

$$\|Ax\| \approx \|(-A^2)^{1/2} x\| \qquad \text{for } x \in \mathcal{D}(A) = \mathcal{D}((-A^2)^{1/2}),$$

i.e. to the boundedness and invertibility of $A(-A^2)^{-1/2}$ which equals the Hilbert transform if $A = \frac{d}{dx}$ on $L_p(\mathbb{R})$. In the case of Hilbert space and operators A with a bounded H^∞-calculus this was first shown in [14], see also [72] for the Banach space case.

If $A_1, \ldots, A_n$ is a commuting family of 1-bisectorial operators, the classical example $A_i = \frac{d}{dx_i}$, $i = 1, \ldots, n$, on $L_p(\mathbb{R}^n)$ motivates us to call the operators $A_i(-A)^{-1/2}$, where $A = \sum_{i=1}^n A_i^2$, Riesz transforms for $A_1, \ldots, A_n$. If $A_1, \ldots, A_n$ have a bounded joint H^∞-calculus, one can not only show that these Riesz transforms are bounded, but obtains inequalities of the form

$$\mathbb{E}\left\| \sum_{i=1}^n r_i A_i x \right\| \approx \left\| (-A)^{1/2} x \right\|, \qquad \mathbb{E}\left\| \sum_{i,j=1}^n r_{ij} A_i A_j x \right\| \approx \|Ax\|,$$

where $(r_i), (r_{ij})$ are independent $\{-1, 1\}$-distributed Bernoulli random variables. These extend the classical inequalities

$$\left\| \left(\sum_{i=1}^n \left| \frac{\partial}{\partial x_i} f \right|^2 \right)^{1/2} \right\|_{L_p} \approx \left\| (-\Delta)^{1/2} f \right\|_{L_p}, \qquad \left\| \left(\sum_{i,j=1}^n \left| \frac{\partial^2}{\partial x_i \partial x_j} f \right|^2 \right)^{1/2} \right\|_{L_p} \approx \|\Delta f\|_{L_p}$$

for $f \in L_p(\mathbb{R}^n)$ with $1 < p < \infty$. This is the content of Section 4.

For bisectorial invertible operators that generate a bounded group, it is contained implicitly in [97] that the boundedness of the spectral projection is equivalent to the boundedness of the Hilbert transform of the group. The argument involves the *analytic generator* of the group. We will give details on how to derive the boundedness of the spectral projection. Then we sketch an alternative more direct approach to this result using the H^∞-functional calculus. In this way we can avoid the assumption $0 \in \rho(A)$.

2.2   The case of bisectorial operators

We will begin to study the special case of a bisectorial operator, from which we will deduce the general case by iteration.

Let $A \in \mathrm{Sect}(\alpha, \omega)$ with $\alpha \in (-\pi, \pi]^2$ and $\omega < \omega' \in [0, \pi)^2$ admissible. We assume that A is densely defined and has dense range. Recall that this implies that A is injective.

We will write $\Sigma = \Sigma_{\alpha, \omega}$, $\Sigma' = \Sigma_{\alpha, \omega'}$, $\Sigma_i = \Sigma_{\alpha_i, \omega_i}$ and $\Sigma_i' = \Sigma_{\alpha_i, \omega_i'}$ ($i = 1, 2$) shortly. We say that $X = X_1 \oplus X_2$ is a *spectral decomposition* of X if X_i is invariant under A and if the restriction A_i of A to X_i is in $\mathrm{Sect}(\alpha_i, \omega_i)$. In particular, $\sigma(A_i) = \sigma(A) \cap \overline{\Sigma_i}$. Recall that we say that the set $Y \subset X$ is *invariant under* A if $Ay \in Y$ for all $y \in \mathcal{D}(A) \cap Y$. Sometimes we will call A_i a *reduced operator*.

The following proposition states uniqueness of such a decomposition and links it to the functional calculus. Clearly, if the operator A would not be injective, there could be no hope to obtain uniqueness as we could choose to incorporate the kernel of A into either of the two subspaces.

2.2.1 Proposition. *Let $X = X_1 \oplus X_2$ be a spectral decomposition of X and denote by P the canonical projection onto X_1, that is $P : (x = x_1 \oplus x_2 \mapsto x_1) : X \to X_1$. Then $P = p(A)$, where $p \in H^\infty(\Sigma')$ is the indicator function of the set Σ_1'.*

Proof. Since $R(z, A_i) = R(z, A)|_{X_i}$ for all $z \in \rho(A) = \rho(A_1) \cap \rho(A_2)$, it is easy to see that

$$f(A) = f(A_1) \oplus f(A_2) \tag{2.1}$$

with $\mathcal{D}(f(A)) = \mathcal{D}(f(A_1)) \oplus \mathcal{D}(f(A_2))$ for all $f \in H^\infty(\Sigma')$. The fact that $A_i \in \text{Sect}_d(\alpha_i, \omega_i)$ together with the consistency of the functional calculus (see Remark 1.3.5) implies that the $H^\infty(\Sigma_i')$-functional calculus for A_i extends its $H^\infty(\Sigma')$-functional calculus. As p is constant on both Σ_1' and Σ_2' we obtain $p(A) = p(A_1) \oplus p(A_2) = I_{X_1} \oplus 0 = P$. $\qquad\square$

In the following we will see that the boundedness of $p(A)$ already induces a spectral decomposition. It follows readily from elementary properties of the functional calculus that $P := p(A)$ is then a bounded projection commuting with A. Thus $X_1 = \mathcal{R}(P)$ and $X_2 = \ker P$ are A-invariant subspaces of X. We denote by A_i the restriction of A to X_i. It remains to show that $A_i \in \text{Sect}(\alpha_i, \omega_i)$.

First, we verify that the spectrum of A_i is situated where it should be.

2.2.2 Proposition. *Using the notation above: If P is bounded then $\sigma(A_i) = \sigma(A) \cap \overline{\Sigma_i}$.*

Proof. It suffices to consider the case $i = 1$ ($i = 2$ follows by symmetry, as $p(A)$ is bounded if and only if $(1 - p)(A)$ is bounded) and to show that $0 \neq \mu \in \overline{\Sigma_2}$ is in the resolvent set of A_1. Fix $\lambda \in \mathbb{C} \setminus \overline{\Sigma'}$. Then the identity

$$(\mu - z)^{-1} = \lambda\mu^{-1}(\lambda - z)^{-1} + (\lambda - \mu)\mu^{-1}z(\lambda - z)^{-1}(\mu - z)^{-1}$$

inserted into the functional calculus together with Corollary 1.4.13 yields

$$(r_\mu p)(A) = (c_1 r_\lambda p + c_2 hp)(A) = c_1 R(\lambda, A)P + c_2 (hp)(A), \tag{2.2}$$

where $h(z) = z(\lambda - z)^{-1}(\mu - z)^{-1}$, $r_c(z) = (c - z)^{-1}$, $c_1 = \lambda\mu^{-1}$ and $c_2 = (\lambda - \mu)\mu^{-1}$. As $hp \in H_0^\infty(\Sigma')$, the operator $(hp)(A)$ is bounded and hence also $(r_\mu p)(A)$. The identity $(\mu - A)(r_\mu p)(A) = p(A)$ and the fact that A and $(r_\mu p)(A)$ commute imply that $\mu \in \rho(A_1)$ and $R(\mu, A_1) = (r_\mu p)(A)|_{X_1}$. $\qquad\square$

In the next step we will establish the bound on the resolvent.

2.2.3 Proposition. *Using the notation above: If P is bounded, then $A_i \in \text{Sect}(\alpha_i, \omega_i)$.*

Proof. By (2.2) we have

$$\|\mu R(\mu, A_1)\| \leq \|P\| \, \|\mu(r_\mu p)(A)\| \leq \|P\| \left(\|\lambda R(\lambda, A)\| + |\lambda - \mu| \, \|(hp)(A)\| \right),$$

where $0 \neq \mu \in \overline{\Sigma_2}$ and λ is chosen in $\mathbb{C} \setminus \overline{\Sigma'}$ with $|\lambda| = |\mu|$. Then the first term is bounded by our sectoriality assumption. Since $|\lambda - \mu| \leq 2\,|\mu|$, we only have to find an upper bound for $|\mu| \, \|(hp)(A)\|$, $0 \neq \mu \in \Sigma_2'$. A direct estimate gives

$$\|(hp)(A)\| \leq C \int_{\Gamma_+} \frac{|dz|}{|\lambda - z| \, |\mu - z|},$$

where $\Gamma_+ = \partial\Sigma_{\alpha,\omega''} \cap \Sigma_1'$, for some $\omega < \omega'' < \omega'$.

We split the contour Γ_+ into two parts: $\Gamma_+^1 = \Gamma_+ \cap B_{2|\mu|}$ and $\Gamma_+^2 = \Gamma_+ \cap B_{2|\mu|}^c$. Then we have for $z \in \Gamma_+^1$ the estimates

$$|\lambda - z| \geq \operatorname{dist}(\lambda, \Gamma_+^1) \geq c\,|\lambda|\,,$$

and similarly

$$|\mu - z| \geq \operatorname{dist}(\mu, \Gamma_+^1) \geq c\,|\mu|\,,$$

for some constant $c \in (0,1]$; consider the first case: either the angle between the line segment joining λ and 0 and the nearer branch of Γ_+ is greater than $\pi/2$, then $\operatorname{dist}(\lambda, \Gamma_+) = |\lambda|$ or the angle is $\phi \in (0,\pi]$ then the distance is $|\lambda|\sin\phi$. The second estimate is done similarly. Therefore,

$$\int_{\Gamma_+^1} \frac{|dz|}{|\lambda - z|\,|\mu - z|} \leq c^{-2}\,|\Gamma_+^1|\,|\mu|^{-2} = 4c^{-2}\,|\mu|^{-1}\,.$$

If $z \in \Gamma_+^2$, then $|\lambda - z| \geq |z| - |\lambda| \geq |z|/2$ and analogously, $|\mu - z| \geq |z|/2$. Hence,

$$\int_{\Gamma_+^2} \frac{|dz|}{|\lambda - z|\,|\mu - z|} \leq 4\int_{\Gamma_+^2} \frac{|dz|}{|z|^2} = 8\int_{2|\mu|}^{\infty} t^{-2}\,dt = 4\,|\mu|^{-1}\,.$$

Therefore,

$$|\lambda - \mu|\,\|(hp)(A)\| \leq 2\,|\mu|\,\|(hp)(A)\| \leq 8(c^{-2} + 1),$$

which proves the claim for $i = 1$. The resolvent estimate for A_2 follows by symmetry, as $p(A)$ is bounded if and only if $(1 - p)(A)$ is bounded. We give a different proof for the estimate in [53]. $\qquad\square$

2.2.4 Remarks.   1. If A has a bounded $H^\infty(\Sigma')$-functional calculus, it is easy to see that the spectral projection is bounded and induces a spectral decomposition.

2. If A is invertible, we need no decay in 0 and the function r_μ can be inserted into the functional calculus without the detour made above. However, as Example 1.9.7 shows, there are bisectorial operators having a bounded sectorial functional calculus and allowing a bounded spectral projection which are neither invertible nor have a bounded bisectorial functional calculus.

Unbounded projection.   Next, given some Banach space admitting a Schauder basis, we will construct an example of an invertible bisectorial operator $A \in \operatorname{Sect}_d(\alpha,\omega)$ with $\alpha = (-\pi/2, \pi/2)$ and $\omega = (0,0)$ whose spectral projection is unbounded. Another example can be found in [108, Ex. 1.4.3] and [91].

Given a Banach space X, we say that a sequence $(e_n)_{n\in\mathbb{N}}$ is a *Schauder basis*, or basis for short, if for every $x \in X$ there is a unique sequence $(x_n)_n \subset \mathbb{C}$ such that $x = \sum_{n=1}^{\infty} x_n e_n$ in X. If this series converges unconditionally we call the basis *unconditional* otherwise *conditional*. A basis is conditional if and only if there is a sequence of signs $(\theta_n)_n \in$

$\{-1, 1\}^{\mathbb{N}}$ and an element $x \in X$ such that $x = \sum_{n=1}^{\infty} x_n e_n$ but the series $x = \sum_{n=1}^{\infty} \theta_n x_n e_n$ does not converge, see [87, Chapter 1], [44] or [51, Chapter 1]. The unit vector bases for ℓ_p, $1 \le p < \infty$, are examples of unconditional bases. The summing basis in c_0 is an example of a conditional basis. We note that each Banach space admitting a basis admits also a conditional basis, see [112, Thm. 23.2].

If $(e_n)_n$ is a Schauder basis of X, we define the projections

$$P_n : \sum_{k=1}^{\infty} x_k e_k \mapsto \sum_{k=1}^{n} x_k e_k.$$

It is easy to see that P_n is bounded and, by the uniform boundedness principle, the number $M_0 = \sup_n \|P_n\|$ is finite. This number is called the *basis constant* of the basis (e_n). Given a scalar sequence $a = (a_n)_n$ we define its variation norm by

$$|a| = \limsup_n |a_n| + \sum_{n \in \mathbb{N}} |a_{n+1} - a_n|,$$

which may be infinite. With a we associate a multiplication operator A on X by setting

$$\mathcal{D}(A) = \{x = \sum_{n=1}^{\infty} x_n e_n \in X : \sum_{n=1}^{\infty} a_n x_n e_n \text{ converges}\}$$

and $Ax = \sum_{n=1}^{\infty} a_n x_n e_n$ for $x \in \mathcal{D}(A)$. It is easy to see that A is a closed and densely defined operator. Partial summation yields that A is bounded if $|a| < \infty$ and in this case we have $\|A\| \le M_0 |a|$, see [118], [51, Section 3.3].

2.2.5 Example. Let X be a Banach space with conditional basis $(e_n)_n$. This is equivalent to the existence of a set $\mathcal{J} \subset \mathbb{N}$ such that the projection $P_{\mathcal{J}} : \mathcal{D} := \langle e_n : n \in \mathbb{N} \rangle \to X$ defined by

$$P_{\mathcal{J}} : x = \sum_{k} x_k e_k \mapsto \sum_{k \in \mathcal{J}} x_k e_k$$

is unbounded. Define $a_n = i2^n$ if $n \in \mathcal{J}$ and $a_n = -i2^n$ otherwise. Denote by A the multiplication operator associated to that sequence.

The operator A is bisectorial, more precisely, $A \in \mathrm{Sect}_d(\alpha, \omega)$ with $\alpha = (-\pi/2, \pi/2)$ and $\omega = (0, 0)$. To prove that $\lambda \in \rho(A)$ and to obtain the required resolvent bound, we have to estimate $|\lambda r_\lambda|$, where r_λ denotes the scalar sequence $((\lambda - a_n)^{-1})$. A direct computation yields the desired result.

Let $f \in H^{\infty}(\Sigma')$, where $\Sigma' = \Sigma_{\alpha, \omega'}$ with $\omega' > 0$. As the projections P_n commute with A, it follows easily that

$$f(A) \left(\sum_{k=1}^{n} x_n e_n \right) = \sum_{k=1}^{n} f(a_n) x_n e_n$$

for each finitely supported $x \in \mathcal{D} \subset \mathcal{D}(A)$. Therefore, $f(A)$ is bounded if and only if $\sum_{k=1}^{\infty} f(a_n) x_n e_n$ converges if $\sum_{k=1}^{\infty} x_n e_n$ does. Hence, the spectral projection $P = p(A)$, where p denotes the indicator function of the upper complex half-plane, is given by the closure of $P_{\mathcal{J}}$, which is not bounded.

2.2.6 Remarks. Let $A \in \mathrm{Sect_d}(\alpha, \omega)$ be bisectorial. If p denotes the indicator function of the set Σ_1', we denote the spectral projections by $P = p(A)$ and $Q = I - P = (1-p)(A)$. The elementary properties of the functional calculus imply readily that P and Q are closed, densely defined operators and that

1. $\mathcal{D}(A) \cap \mathcal{R}(A) \subset \mathcal{D}(P) = \mathcal{D}(Q) =: D$;

2. $P + Q = I|_D$;

3. $P^2 = P$ and $Q^2 = Q$ (with equality of domains);

4. $PQ = QP = 0|_D$.

The subspaces $X_1 = \mathcal{R}(P)$ and $X_2 = \mathcal{R}(Q) = \ker P$ are A-invariant and the sum $X_1 \oplus X_2$ is direct; in general however, neither is this sum equal to X nor are the subspaces X_i complemented.

We note the following corollary of Theorem 1.10.12 and Remark 1.10.15.

2.2.7 Corollary. Let $A \in \mathrm{aRSect_d}(\alpha, \omega)$ with $\alpha \in (-\pi, \pi]^2$ and $\omega \in [0, \pi)^2$ admissible. If $e^{i\theta} A$ has a bounded sectorial functional calculus for some suitable θ, then the spectral projection P is bounded.

2.2.8 Remarks.   1. If A is an invertible bisectorial operator, it follows from Dore's theorem (Thm. 1.8.3) that the spectral projection is always bounded in the real interpolation spaces $(X, \mathcal{D}(A))_{\theta,p}$ for all $\theta \in (0,1)$ and all $p \in [1, \infty]$. This result goes back to Grisvard [60] where the author used it to deduce the boundedness of P in the Hilbert space setting assuming equality of the interpolation spaces $(X, \mathcal{D}(A))_{\theta,2}$ and $(X, \mathcal{D}(A^*))_{\theta,2}$ for some $\theta \in (0,1)$. This condition is related to BIP, see [125].

2. A subset $\sigma_1 \subset \sigma(A)$ is called a *spectral set* if both σ_1 and $\sigma(A) \setminus \sigma_1$ are closed in $\mathbb{C}$. If σ_1 is a bounded spectral set, one can define a bounded projection by defining $\tilde{P} = (2\pi i)^{-1} \int_\gamma R(z, A)\, dz$, where γ is a suitable cycle of Jordan curves surrounding the spectrum counterclockwise. Then $\tilde{P}$ induces a decomposition of the Banach space X such that $\sigma(A_1) = \sigma_1$ and $\sigma(A_2) = \sigma(A) \setminus \sigma_1$ (see [89, Section A.1]). However, the operator A_2 won't be sectorial, in general. It is easy to construct a counterexample ([108, Ex. 1.3.8]) by means of the generator A_J of the Riemann-Liouville semigroup (see [68, p.663] or [7]). Recall that $-A_J$ is sectorial of angle 0 and has empty spectrum.

However, if A (or $-A$) generates an analytic C_0-semigroup such that $\sigma(A) \cap i\mathbb{R} = \emptyset$, then the part of $\sigma(A)$ contained in the right (resp. left) half-plane is a spectral set and the projection $\tilde{P}$ defined above agrees with the spectral projection P defined by means of the functional calculus. The reason is that the analyticity of the semigroup implies the desired sectoriality estimate (see [89], [108, Ex. 1.4.1, 1.4.2]).

Spectral projection and fractional powers. Now we will link the notions of fractional powers and the spectral projections. Let $0 \le \theta_1 < \theta_2 < 2\pi$ and $w \in \mathbb{C}$. We will study the function

$$h = h_{\theta_1,\theta_2,w} : z \mapsto (e^{i\theta_1}z)^w (e^{i\theta_2}z)^{-w},$$

which is defined on $\mathbb{C}\backslash\gamma$ for $\gamma = e^{i(\pi-\theta_1)}[0,\infty) \oplus e^{i(\pi-\theta_2)}(0,\infty)$. Note that, if we interchange θ_1 and θ_2, this corresponds to substituting w by $-w$. As the sum of the exponents is zero, we have, for $z \notin \gamma$,

$$h(z) = e^{iw(\arg(e^{i\theta_1}z) - \arg(e^{i\theta_2}z))},$$

where $\arg : \mathbb{C} \setminus (-\infty, 0] \to (-\pi, \pi]$ denotes the argument function associated with the principal branch of the logarithm used to define the complex powers z^w. As for $\theta \in [0, 2\pi)$

$$\arg(e^{i\theta}z) = \begin{cases} \theta + \arg(z) & \text{if } \arg z \in (-\pi, \pi - \theta], \\ \theta + \arg(z) - 2\pi & \text{if } \arg z \in (\pi - \theta, \pi], \end{cases}$$

we have

$$h(z) = \begin{cases} e^{iw(\theta_1-\theta_2)} & \text{if } \arg(z) \in (-\pi, \pi - \theta_2] \cup (\pi - \theta_1, \pi], \\ e^{iw(\theta_1-\theta_2+2\pi)} & \text{if } \arg(z) \in (\pi - \theta_2, \pi - \theta_1). \end{cases} \qquad (2.3)$$

That is, if $w \notin \mathbb{Z}$, the function h takes two different values. We will use this fact to relate h and the indicator function of the set Σ_1'. Recall that we assume in this section that $A \in \text{Sect}(\alpha, \omega)$, with $\alpha \in (-\pi, \pi]^2$ and $\omega < \omega' \in [0, \pi)^2$ admissible, is densely defined with dense range. Writing $\Sigma = \Sigma_{\alpha,\omega}$, $\Sigma' = \Sigma_{\alpha,\omega'}$, $\Sigma_i = \Sigma_{\alpha_i,\omega_i}$ and $\Sigma_i' = \Sigma_{\alpha_i,\omega_i'}$, we choose $\eta_k \in (-\pi, \pi]$ such that each connected component of $\mathbb{C} \setminus \overline{\Sigma'}$ contains one of the points $e^{i\eta_k}$. We put $\theta_1 = \min\{\pi - \eta_1, \pi - \eta_2\}$ and $\theta_2 = \max\{\pi - \eta_1, \pi - \eta_2\}$. This implies $0 \le \theta_1 < \theta_2 < 2\pi$ and the function $h = h_{\theta_1,\theta_2,w}$ is in $H^\infty(\Sigma')$ and is constant on each component of Σ', say $h|_{\Sigma_i'} = c_i \ne 0$. Therefore, if $w \notin \mathbb{Z}$, then $h = c_1 p + c_2(1-p)$ or equivalently

$$p = (c_1 - c_2)^{-1}(h - c_2).$$

Applying the H^∞-functional calculus to this identity yields immediately

$$P = p(A) = (c_1 - c_2)^{-1}(h - c_2)(A) = (c_1 - c_2)^{-1}h(A) - c_2(c_1 - c_2)^{-1}, \qquad (2.4)$$

that is, the spectral projection is bounded if and only if $h(A)$ is bounded.

2.2.9 Theorem. *Using the notation of the preceding paragraph, let $A \in \text{Sect}_d(\alpha, \omega)$. The following statements are equivalent:*

1. *$P = p(A)$ is bounded.*

2. *There exists $w \in \mathbb{C} \setminus \mathbb{Z}$ such that $(e^{i(\pi-\eta_1)}A)^w(e^{i(\pi-\eta_2)}A)^{-w}$ has a bounded extension to all of X.*

3. *For all $w \in \mathbb{C} \setminus \mathbb{Z}$ the operator $(e^{i(\pi-\eta_1)}A)^w(e^{i(\pi-\eta_2)}A)^{-w}$ admits a bounded extension to all of X.*

2.2.10 Remark. For invertible 1-bisectorial operators these equivalences are already shown in [49] with a different proof. The use of the functional calculus makes the arguments more transparent and allows to discard the invertibility assumption.

Proof. This is immediate from the identity (2.4) and the fact that the densely defined operator $(e^{i(\pi-\eta_1)}A)^w(e^{i(\pi-\eta_2)}A)^{-w}$ is a restriction of $h(A)$, see Theorem 1.4.12 (the operator is densely defined, as its domain contains the range of $\Psi(A)$). $\qquad\square$

Instead of speaking of a bounded extension, we will often just say that the operator $(e^{i(\pi-\eta_1)}A)^w(e^{i(\pi-\eta_2)}A)^{-w}$ is bounded without implying that its domain is actually equal to X. Note that this is equivalent to the following equivalence of norms:

$$\left\|(e^{i(\pi-\eta_1)}A)^w x\right\| \sim \left\|(e^{i(\pi-\eta_2)}A)^w x\right\|, \qquad x \in D$$

where $D = \mathcal{D}((e^{i(\pi-\eta_j)}A)^w)$ is independent of $j \in \{1,2\}$.

Consequently, a 1-bisectorial operator A admitting bounded imaginary powers has a bounded spectral projection if and only if $-A$ also admits bounded imaginary powers. In Hilbert space it suffices to know that A has bounded imaginary powers. We do not know whether this remains true in general Banach spaces.

Combining this fact with the characterization of the boundedness of the functional calculus found above we obtain an example of a bisectorial operator with BIP-type greater or equal to π.

An operator with BIP-type greater or equal to π. Recall that in Example 1.9.5 we constructed a densely defined bisectorial operator B with dense range on a UMD space; it admits bounded imaginary powers (with type $< \pi$), because it admits even a bounded sectorial functional calculus. Since, by construction, it has a bounded spectral projection, we know that $-B$ does also admit bounded imaginary powers (by Theorem 2.2.9). However, the BIP-type of $-B$ is not less than π. Indeed, if it was less than π, the operator B would be almost R-bisectorial (by Proposition 1.10.9), and hence B would admit a bounded bisectorial functional calculus (by Theorem 1.10.12), which does not hold. We summarize these considerations.

2.2.11 Proposition. *There is a UMD space X and a densely defined bisectorial operator A with dense range defined on it that has bounded imaginary powers with BIP-type $\theta_A \geq \pi$. Moreover, the operator $-A$ has a bounded functional calculus.*

An operator with similar properties has already been discovered by Haase (see [61, Chapter 3 §5], [62]); although by different means. His approach is based on a representation theorem for C_0-groups (of group type less than π) due to Monniaux.

Observe that, although B has no bounded bisectorial functional calculus, B has a bounded spectral projection. Hence, $f(B)$ is bounded for some, but not all, bounded holomorphic functions.

In the following we will characterize the boundedness of the spectral projection in terms of the domains of the fractional powers $(e^{i(\pi-\eta_k)}A)^w$; however we will assume that the operator A is invertible.

2.2.12 Theorem. *In addition to the assumptions in Thm. 2.2.9 let A be invertible. Then the following statements are equivalent:*

1. *$P = p(A)$ is bounded.*

2. *There exists $w \in \mathbb{C}\backslash\mathbb{Z}$ with $\mathcal{R}e(w) > 0$ such that $\mathcal{D}((e^{i(\pi-\eta_1)}A)^w) \supset \mathcal{D}((e^{i(\pi-\eta_2)}A)^w)$.*

3. *For all $w \in \mathbb{C}$ with $\mathcal{R}e(w) > 0$ we have $\mathcal{D}((e^{i(\pi-\eta_1)}A)^w) = \mathcal{D}((e^{i(\pi-\eta_2)}A)^w)$.*

Proof. The implication 3. $\Rightarrow$ 2. is trivial. The implication 2. $\Rightarrow$ 1. is true, as the relation $\mathcal{R}((e^{i(\pi-\eta_2)}A)^{-w}) = \mathcal{D}((e^{i(\pi-\eta_2)}A)^w)$ and the fact that the invertibility of A implies the boundedness of $(e^{i(\pi-\eta_2)}A)^{-w}$ tell us $\mathcal{D}(h(A)) \supset \mathcal{D}((e^{i(\pi-\eta_1)}A)^w(e^{i(\pi-\eta_2)}A)^{-w}) = X$. Finally, we show 1. $\Rightarrow$ 3.: Let P be bounded and denote by $A_k \in \text{Sect}(\alpha_k, \omega_k)$ the reduced operators, then $\mathcal{D}((e^{i(\pi-\eta_k)}A)^w) = \mathcal{D}((e^{i(\pi-\eta_k)}A_1)^w) \oplus \mathcal{D}((e^{i(\pi-\eta_k)}A_2)^w)$ (compare (2.1)). Now it suffices to observe that, as h is constant on Σ'_k, we have $h(A_k) = c_k \neq 0$, hence $(e^{i\theta_2}A_k)^{-w} = c_k(e^{i\theta_1}A_k)^{-w}$; in particular, the ranges agree, which proves the claim. $\square$

We conclude this section with a last characterization of the boundedness of the spectral projection in terms of the logarithm. Note, the function log is in $H_P(\Sigma_\omega)$ for $\omega \in (0, \pi)$ and hence $z \mapsto \log(e^{i(\pi-\eta_k)}z) \in H_P(\Sigma')$. We have the following identity for $z \in \Sigma'$

$$l(z) := \log(e^{i\theta_1}z) - \log(e^{i\theta_2}z) = i(\arg(e^{i\theta_1}z) - \arg(e^{i\theta_2}z));$$

observe that this function is constant on both Σ'_1 and Σ'_2 taking as value either $i(\theta_1 - \theta_2)$ or $i(\theta_1 - \theta_2 + 2\pi)$. Therefore, we have

2.2.13 Theorem. *Let $A \in \text{Sect}_d(\alpha, \omega)$. Using the notation above, we have the following equivalent statements:*

1. *$P = p(A)$ is bounded.*

2. *$l(A)$ is bounded.*

3. *$\log(e^{i(\pi-\eta_1)}A) - \log(e^{i(\pi-\eta_2)}A)$ has a bounded extension to X.*

Proof. The proof is essentially the same as the proof of Theorem 2.2.9. $\square$

This result was proven already in [49] for invertible 1-bisectorial operators A by different means.

We close this section considering the spectral decomposition in the setting of asymptotically bisectorial operators.

Asymptotically bisectorial operators. Let $A \in \mathrm{ASect}_\mathrm{d}(\Omega_{(0,\pi),(\omega,\omega)})$ be an asymptotically bisectorial operator as defined in Section 1.7.2. Assume that its spectrum $\sigma(A)$ decomposes into the disjoint union of two sets that are both open and closed in $\sigma(A) = \sigma_1 \cup \sigma_2$ such that $\sigma_1 \cap \{z : |z| > r\} \subset \{z : \mathcal{R}e(z) < 0\}$ and $\sigma_2 \cap \{z : |z| > r\} \subset \{z : \mathcal{R}e(z) > 0\}$ for some $r > 0$. Then there is an admissible set Ω for A that is the disjoint union of two open sets Ω_1 and Ω_2 such that $\sigma_k \subset \Omega_k$, $k = 1, 2$.

Let $p \in H^\infty(\Omega)$ denote the indicator function of the set Ω_1. It is easy to verify that the Banach space X admits a unique spectral decomposition $X = X_1 \oplus X_2$ such that the spaces X_i are A-invariant and the restriction A_i of A to X_i is quasi-sectorial with spectrum $\sigma(A_i) = \sigma_i$ if and only if $p(A)$ is a bounded operator. The relevant propositions (Prop. 2.2.2, 2.2.3) may be adapted easily to this situation.

If A has a bounded $H^\infty(\Omega)$-functional calculus, then there is a spectral decomposition.

2.2.14 Proposition. *Let $A \in \mathrm{ASect}_\mathrm{d}(\Omega_{(0,\pi),(\omega,\omega)})$ be an asymptotically bisectorial operator and suppose that its spectrum $\sigma(A)$ decomposes into the disjoint union of two sets that are both open and closed in $\sigma(A) = \sigma_1 \cup \sigma_2$ such that $\sigma_1 \cap \{z : |z| > r\} \subset \{z : \mathcal{R}e(z) < 0\}$ and $\sigma_2 \cap \{z : |z| > r\} \subset \{z : \mathcal{R}e(z) > 0\}$, $r > 0$. If $iA, -iA \in QH^\infty$, then there is a spectral decomposition.*

Proof. By Corollary 1.7.19 A has a bounded $H^\infty(\Omega)$-functional calculus for all admissible sets Ω having sufficiently large opening angle. We may find an admissible set Ω having a large enough opening angle that is the disjoint union of two open sets Ω_1 and Ω_2 such that $\sigma_k \subset \Omega_k$, $k = 1, 2$. For $p = 1_{\Omega_1} \in H^\infty(\Omega)$ the spectral projection is given by $p(A) \in \mathcal{L}(X)$. □

The easiest example of operators satisfying this spectral separation condition are of course invertible bisectorial operators. We formulate this special case in the following corollary.

2.2.15 Corollary. *Let $A \in \mathrm{Sect}_\mathrm{d}(\Sigma_{(-\pi/2,\pi/2),(\omega,\omega)})$ be a densely defined invertible bisectorial operator on a Banach space. If both A and $-A$ are in QH^∞, then there is a spectral decomposition.*

2.3   Boundedness of Riesz transforms

The results of this section are motivated by the following well known example.

2.3.1 Example. Let $p \in [1, \infty)$ and define $A = \frac{d}{dt}$ on $L_p(\mathbb{R}, X)$ with domain $\mathcal{D}(A) = W^{1,p}(\mathbb{R}, X)$. As A is the generator of the bounded group of left translations on $L_p(\mathbb{R}, X)$, it follows that $A \in \mathrm{Sect}_\mathrm{d}(\alpha, \omega)$ with $\alpha = (-\pi/2, \pi/2)$ and $\omega = (0, 0)$. Indeed, the resolvent is given by the Laplace transform of the contractive semigroup of left translations; hence $\|R(z, A)\| \leq |\mathcal{R}e(z)|^{-1}$ for all $z \in \mathbb{C} \setminus i\mathbb{R}$. In order to decide whether the spectral projection corresponding to the upper complex half-plane is bounded we compute the operator $A^{1/2}(-A)^{-1/2}$ using Fourier multipliers.

Since A is the Fourier multiplier associated to the symbol $m(x) = ix$, the operators $A^{1/2}$ and $(-A)^{-1/2}$ have the symbols (compare Example 1.5.9)

$$(ix)^{1/2} = e^{1/2(\log|x| + i\arg(ix))} \text{ and } (-ix)^{-1/2} = e^{-1/2(\log|x| + i\arg(-ix))}.$$

The symbol of the operator $A^{1/2}(-A)^{-1/2}$ is given as the product of the symbols computed above, hence by (2.3)

$$(ix)^{1/2}(-ix)^{-1/2} = e^{i/2(\arg(ix) - \arg(-ix))} = i \operatorname{sign}(x).$$

Thus, up to a constant $A^{1/2}(-A)^{-1/2}$ is the Hilbert transform which is well known to be bounded, or more precisely to admit a bounded extension, if and only if $p \in (1, \infty)$ and if X is a UMD space [119]. The spectral projection P associated to A corresponding to the upper half-plane has symbol $1_{\mathbb{R}_+}$.

Observe that the boundedness of $A^{1/2}(-A)^{-1/2}$ corresponds to the following norm inequality: $\left\| A^{1/2}x \right\| \leq C \left\| (-A)^{1/2}x \right\|$ for all $x \in \mathcal{D}(A^{1/2}) \cap \mathcal{D}((-A)^{1/2})$. Note that $A^{1/2}(-A)^{-1/2}$ is bounded if and only if $A^{-1/2}(-A)^{1/2}$ is bounded. These considerations imply the equivalence of the following statements:

1. X is a UMD space and $p \in (1, \infty)$;

2. the spectral projection P (the Poisson projection) is bounded on $L_p(\mathbb{R}, X)$;

3. the operator $\left(\frac{d}{dx}\right)^{1/2}\left(-\frac{d}{dx}\right)^{-1/2}$ is bounded on $L_p(\mathbb{R}, X)$;

4. $\left\| \left(\frac{d}{dx}\right)^{1/2}f \right\| \approx \left\| \left(-\frac{d}{dx}\right)^{1/2}f \right\|$ $\quad \forall f \in \mathcal{D}(\left(\frac{d}{dx}\right)^{1/2}) \cap \mathcal{D}(\left(-\frac{d}{dx}\right)^{1/2})$.

By this equivalence of norms we obtain that the graph norms $\|x\|_{A^{1/2}} = \left\| A^{1/2}x \right\| + \|x\|$ and $\|x\|_{(-A)^{1/2}} = \left\| (-A)^{1/2}x \right\| + \|x\|$ are equivalent. As both domains contain the common core $\mathcal{D}(A) \cap \mathcal{R}(A)$ (see Remark 1.4.19) the closures agree, that is $\mathcal{D}(A^{1/2}) = \mathcal{D}((-A)^{1/2})$.

Remark that, if X is a UMD space and $p \in (1, \infty)$, we know that A has a bounded $H^\infty(\Sigma_{\alpha, \omega'})$-functional calculus for all $\omega' > 0$, see Example 1.5.9. Moreover, the UMD property of X is necessary for A to possess a bounded functional calculus. Actually, it is known that 1. is equivalent to A admitting BIP [102, p. 216].

Let us consider a similar example with the Laplacian $\Delta = \left(\frac{d}{dt}\right)^2$ on $L_p(\mathbb{R}, X)$.

2.3.2 Example. Let $p \in [1, \infty)$ and let $A = \frac{d}{dt}$ with domain $\mathcal{D}(A) = W^{1,p}(\mathbb{R}, X)$ as in the previous example. Consider the function $h : \mathbb{C} \setminus \mathbb{R} \to \mathbb{C}$, $h(z) = (-z^2)^{1/2}z^{-1}$. If $\arg(z) \in (0, \pi)$ and $z = re^{i\arg(z)}$, then $z^2 = r^2 e^{i2\arg(z)}$, so $-z^2 = r^2 e^{i(2\arg(z) - \pi)}$, where $2\arg(z) - \pi \in (-\pi, \pi)$. Hence $(-z^2)^{1/2} = re^{i\arg(z)}e^{-i\pi/2}$ and thus $h(z) = e^{-i\pi/2} = -i$. Similarly, we find that $h(z) = i$ for $\arg(z) \in (-\pi, 0)$. Therefore, the spectral projection is bounded if and only if $h(A)$ is bounded, which is equivalent to the boundedness of $(-A^2)^{1/2}A^{-1}$; here we used the composition rule (Prop. 1.5.7). Note that the boundedness of $(-A^2)^{1/2}A^{-1}$ is equivalent to the boundedness of $A(-A^2)^{-1/2}$. Therefore, we can add the following two equivalences to our list:

5. the operator $\left(-\frac{d^2}{dx^2}\right)^{1/2}\left(\frac{d}{dx}\right)^{-1}$, the Hilbert transform, is bounded;

6. $\left\|\left(-\frac{d^2}{dx^2}\right)^{1/2}f\right\| \approx \left\|\frac{d}{dx}f\right\|$ $\forall f \in W^{1,p}(\mathbb{R}, X)$.

By this equivalence of norms we obtain that the graph norm $\left\|(-A^2)^{1/2}x\right\| + \|x\|$ is equivalent to the Sobolev norm $\|Ax\| + \|x\|$. As both domains contain the common core $\mathcal{D}(A^2) \cap \mathcal{R}(A^2)$ (see Remark 1.4.19), the closures agree, that is $\mathcal{D}((-A^2)^{1/2}) = \mathcal{D}(A) = W^{1,p}(\mathbb{R}, X)$.

With the same arguments as in the proof of Theorem 2.2.9 we can generalize the equivalence 1. $\Leftrightarrow$ 5. Note that for a bisectorial operator $A \in \text{Sect}((-\pi/2, \pi/2), (\theta, \theta))$, the operator $-A^2$ is in $\text{Sect}(2\theta)$, $\theta \in [0, \pi/2)$.

2.3.3 Theorem. *Let A be a bisectorial operator on X. Then the spectral projection P corresponding to the upper half plane is bounded if and only if $A(-A^2)^{-1/2}$ extends to a bounded operator, i.e.*

$$\left\|(-A^2)^{1/2}x\right\| \approx \|Ax\| \text{for } x \in \mathcal{D}(A) = \mathcal{D}((-A^2)^{1/2}).$$

For the Hilbert space case it was noted in [14] that the boundedness of the H^∞-calculus implies the above estimate. See also [72] for the Banach space case.

Let $A_1, \ldots, A_n$ now be a family of commuting bisectorial operators on X. Motivated by the example $A_i = \frac{d}{dx_i}$ on $L_p(\mathbb{R}^n)$ and the classical Riesz transforms $\frac{d}{dx_i}(-\Delta)^{-1/2}$ we call the operators $A_i(-A)^{-1/2}$ with $A = \sum_{i=1}^n A_i^2$ *Riesz transforms* of the family $A_1, \ldots, A_n$. By means of the H^∞-calculus we obtain the boundedness of such Riesz-transforms and related inequalities.

2.3.4 Theorem. *Let $A_1, \ldots, A_n$ be commuting bisectorial operators with a bounded joint $H^\infty(\Sigma^n)$-calculus, where $\Sigma = \Sigma_{(-\pi/2, \pi/2), (\omega, \omega)}$, $\omega \in [0, \pi/4)$. Then we have the inequalities*

$$\mathbb{E}\left\|\sum_{i=1}^n r_i A_i x\right\| \approx \left\|(-A)^{1/2}x\right\|, \qquad \mathbb{E}\left\|\sum_{i,j=1}^n r_{ij} A_i A_j x\right\| \approx \|Ax\|$$

where $(r_i), (r_{ij})$ are independent $\{-1, 1\}$-distributed Bernoulli random variables.

Proof. Consider the holomorphic functions $h(z) = \left(\sum_{i=1}^n \epsilon_i z_i\right)\left(-\sum_{i=1}^n z_i^2\right)^{-1/2}$ and $g(z) = \left(\sum_{i,j=1}^n \epsilon_{ij} z_i z_j\right)\left(\sum_{i=1}^n z_i^2\right)^{-1}$. It is easy to see that $g, h \in H^\infty(\Sigma^n)$. Moreover, the functions are bounded, uniformly with respect to the choice of signs $\epsilon_i, \epsilon_{ij} \in \{-1, 1\}$. Hence, by the boundedness of the joint functional calculus, we obtain the inequalities

$$\left\|\sum_{i=1}^n \epsilon_i A_i x\right\| \leq C \|h\|_\infty \left\|(-A)^{1/2}x\right\|, \qquad \left\|\sum_{i,j=1}^n \epsilon_{ij} A_i A_j x\right\| \leq C \|g\|_\infty \|Ax\|$$

where C denotes the norm of the joint calculus. Observe also that $-A$ is a sectorial operator. Since this estimate is uniform in the choice of signs, we obtain one of the two desired inequalities by integration.

We will obtain the reverse inequalities combining randomization and a duality argument. We will provide the details for the first equivalence, the argument for the second is similar. Denote by $X^{\#}$ the *moon-dual* of X, that is the set $\overline{\mathcal{D}(A^*)} \cap \overline{\mathcal{R}(A^*)}$ where the closure is taken in the norm of X^*. The moon-dual operator $A^{\#}$ of A is the part of A^* in $X^{\#}$; if A is sectorial, densely defined with dense range or has a bounded H^{∞}-calculus then $A^{\#}$ shares these properties. If X is reflexive, we have $A^{\#} = A^*$. We note that $X^{\#}$ is *norming* for X (see [81, Section 15A] for details). Let $x \in \mathcal{D}((-A)^{1/2})$, $y \in \mathcal{D}((-A^{\#})^{1/2})$, then

$$\left| \langle (-\sum A_i^2)^{1/2}x, ((-\sum A_i^2)^{1/2})^{\#}y \rangle \right| = \left| \langle \sum A_i^2 x, y \rangle \right| = \left| \sum \langle A_i x, A_i^{\#} y \rangle \right|$$
$$\leq \left(\mathbb{E} \left\| \sum r_i A_i x \right\|^2 \right)^{1/2} \left(\mathbb{E} \left\| \sum r_i A_i^{\#} y \right\|^2 \right)^{1/2}$$
$$\leq c \mathbb{E} \left\| \sum r_i A_i x \right\| \left\| (-A^{\#})^{1/2} y \right\| ;$$

in the last step we used the Khinchine-Kahane inequality and made use of the inequality proved above (substituting $A^{\#}$ for A). Therefore, since the set $Z = \mathcal{R}((-A^{\#})^{1/2})$ is norming, we find

$$\left\| (-A)^{1/2} x \right\| = \sup_{z \in Z, \|z\|=1} \left| \langle (-A)^{1/2} x, z \rangle \right| \leq c \mathbb{E} \left\| \sum r_i A_i x \right\|$$

which completes the proof. $\qquad\qquad\qquad\qquad\qquad\qquad\qquad\qquad\qquad\qquad\qquad$ $\square$

2.3.5 Example. Let $X = L_p(\mathbb{R}^n, \frac{dx_1}{x_1} \cdots \frac{dx_n}{x_n})$. For $i = 1, \ldots, n$, define on X the semigroup $T_i(t)f(x) = f(x_1, \ldots, x_{i-1}, e^t x_i, x_{i+1}, \ldots, x_n)$. As the measure of the L_p space X is invariant under scaling, we find that T_i are contraction semigroups. Moreover, it is easy to see that they are positive and commute. Their generators, given by $A_i = x_i \frac{d}{dx_i}$, have thus a bounded bisectorial functional calculus (apply the result of Hieber and Prüss [67]). Since the L_p space X has property (α), we find that $(A_1, \ldots, A_n)$ has a bounded joint bisectorial functional calculus.

The operator A_i^2 is just $x_i \frac{d}{dx_i}(x_i \frac{d}{dx_i}) = x_i \frac{d}{dx_i} + x_i^2 \frac{d^2}{dx_i^2}$. Therefore, the preceding theorem gives the equivalence

$$\left\| \left(\sum_{i,j=1}^n \left| x_i x_j \frac{d}{dx_i} \frac{d}{dx_j} f \right|^2 \right)^{1/2} \right\| \approx \left\| \sum_{i=1}^n x_i^2 \frac{d^2}{dx_i^2} f + x_i \frac{d}{dx_i} f \right\| .$$

The operator $B = x\frac{d}{dx} + x^2 \frac{d^2}{dx^2}$ is the Black-Scholes differential operator. For information on the Black-Scholes partial differential equation $u_t = xu_x + x^2 u_{xx}$, $(t, x) \in \mathbb{R} \times (0, \infty)$ see e.g. [13] and [58] and the references therein.

Making use of [81, Cor. 10.15] we may extend this result to the vector valued setting, allowing for functions taking values in a UMD space.

2.4   Bounded C_0-groups

In this section we will study in detail the special case when the bisectorial operator A is actually the generator of a bounded strongly continuous group U. We will characterize the boundedness of the spectral projection in terms of the generalized Hilbert transform.

Let p denote the indicator function of the upper complex half-plane. The example $A = \frac{d}{dt}$ (Example 2.3.1) shows that the spectral projection $P = p(A)$ is bounded if and only if the Hilbert transform is a bounded operator on L_p. In general, let $(U(t))_{t \in \mathbb{R}}$ be a bounded C_0-group on the Banach space X with generator A. For $0 < \epsilon < T < \infty$ consider the operators

$$H^U_{\epsilon,T} x := \frac{i}{\pi} \int_{\epsilon < |s| < T} \frac{U(s)x}{s} \, ds;$$

if the strong limit of these operators (as $\epsilon \to 0$ and $T \to \infty$) exists, it defines a bounded operator H^U which we will call the (generalized) Hilbert transform associated to U. We recover the classical Hilbert transform if U is the group of translations and $X = L_p(\mathbb{R})$.

2.4.1   Approach via the analytic generator

In the following we will study the relation between the spectral projection P and the Hilbert transform H^U. To this end we make use of some results of [97] connecting H^U and the analytic generator of the group U. We state the relevant definitions and results referring to [97] for proofs and details of known facts; the relevant results may also be found in [98].

Denote by Ω_α, $\alpha \in \mathbb{C} \setminus i\mathbb{R}$, the strip $\{z \in \mathbb{C} : \mathcal{R}e(z) \text{ lies between 0 and } \mathcal{R}e(\alpha)\}$. Given a strongly continuous group U on X we define its *analytic extension* $(C_\alpha)_{\alpha \in \mathbb{C}}$ by

$$\mathcal{D}(C_\alpha) = \{x \in X : \exists f_x \in H(\Omega_\alpha, X) \cap C(\overline{\Omega_\alpha}, X), \, f_x(is) = U(s)x, \, s \in \mathbb{R}\}$$

setting $C_\alpha x = f_x(\alpha)$, $x \in \mathcal{D}(C_\alpha)$ for $\alpha \in \mathbb{C} \setminus i\mathbb{R}$ and $C_{is} = U(s)$, $\mathcal{D}(C_{is}) = X$, for $s \in \mathbb{R}$. The operator $C = C_1$ is called the *analytic generator* of the group U. The operators C_α are densely defined and closed; moreover they satisfy the following semigroup property:

$$C_\alpha C_\beta = C_{\alpha+\beta} \qquad \text{if} \quad \mathcal{R}e(\alpha)\,\mathcal{R}e(\beta) > 0.$$

The next theorem gives some information on boundary values of holomorphic semigroups (see [97, p. 15], compare also [12, Prop. 1.1, 1.2]).

2.4.1 Theorem. *Let G be the generator of an analytic C_0-semigroup S of angle $\pi/2$. The following two statements are equivalent:*

1. $\sup\{\|S(z)\| : 0 < \mathcal{R}e(z) < 1, |\mathfrak{Im}(z)| \le 1\} < \infty$;

2. iG generates a strongly continuous group U on X.

If one of the equivalent statements of the theorem is true, we say that S admits a *trace* on $i\mathbb{R}$ or that U is the trace of S. In this case the limit $S(is) := \lim_{t\to 0+} S(t+is)x$ exists for all $x \in X$ and all $s \in \mathbb{R}$ and $U(s) = S(is)$. Moreover, $S(t+is) = S(t)S(is)$ for all $s \in \mathbb{R}$, $t \geq 0$.

Hence, if the group U is the trace of S, we have $C = S(1)$. More is true [97, Cor. 2.17]:

2.4.2 Proposition. *Let C be the analytic generator of the C_0-group U on X. Then C is bounded if and only if U is the trace of an analytic semigroup on X.*

In fact the group $U(s) = C_{is}$ turns out to be the boundary group of the holomorphic semigroup $(C_z)_{\mathcal{R}e(z)>0}$.

Now, we will relate the analytic generator with the Hilbert transform [97, Prop. 3.6]

2.4.3 Proposition. *Let U be a C_0-group. The following assertions are equivalent:*

1. $\rho(C) \neq \emptyset$;

2. $\mathcal{D}(C) + \mathcal{R}(C) = X$;

3. $H_{\epsilon,1}^U$ admits a strong limit as $\epsilon \to 0$.

If the underlying Banach space X has the UMD property and if the group U is bounded, the principal value integral converges for all $x \in X$ and the Hilbert transform H^U is hence a bounded operator [97, Prop. 4.1].

Assume that U is a bounded group and that the Hilbert transform H^U exists as a principal value integral for all $x \in X$ and hence defines a bounded operator on X, then we define $P_0 x = (1 - (H^U)^2)x$, $P_+ x = \frac{1}{2}((H^U)^2 + H^U)x$ and $P_- x = \frac{1}{2}((H^U)^2 - H^U)x$. These are three commuting projections inducing a decomposition of X into U-invariant subspaces $X_0 = \mathcal{R}(P_0)$, $X_+ = \mathcal{R}(P_+)$ and $X_- = \mathcal{R}(P_-)$. It is known that $X_0 \subset \ker A$ [97, Prop. 4.16]. The behaviour of the bounded group U on the other two spaces X_+ and X_- was considered in [97, Théorème 4.19]: denoting by U_+ respectively U_- the part of U in X_+ respectively X_-, we find that $(U_+(s))$ is the trace of a holomorphic semigroup T_+ of angle $\pi/2$ on X_+ and that $(U_-(-s))$ is the trace of a holomorphic semigroup T_- of angle $\pi/2$ on X_-; or, expressed in terms of the analytic generator: the analytic generator C_+ of U_+ is bounded and the analytic generator C_- of U_- is invertible. Following the terminology of [97] we will call $X = X_0 \oplus X_+ \oplus X_-$ the *Hardy space* decomposition of X (with respect to U). This is justified as we recover the classical Hardy spaces by choosing for U the group of translations on $L_p(\mathbb{R}, X)$, $p \in (1, \infty)$ and X UMD [97, Rem. 4.20].

In fact, the semigroup T_+ is given by the analytic extension of U_+; similarly for T_-. Moreover, the analytic extension is identified in [97, Théorème 4.19] as

$$T_+(z)x = \frac{i}{2\pi} \int_{-\infty}^{\infty} \frac{U(s)x}{s + iz} \, ds \qquad (2.5)$$

for $\mathcal{R}e(z) > 0$ and $x \in X_+$. A similar statement holds for T_-, it suffices to replace the group $U(s)$ by $U(-s)$. We claim that these holomorphic semigroups are in effect bounded.

Clearly, it is sufficient to establish the result for T_+. Furthermore, the semigroup property $T_+(s+it) = T_+(s)U_+(t)$, $t \in \mathbb{R}$, $s > 0$, implies that we are left to show the boundedness of the semigroup $(T_+(\delta))_{\delta>0}$ as we assumed the group U to be bounded. In [97, Prop. 4.18] it was showed that $T_+(\delta)$ converges strongly to the identity on X_+ as $\delta \to 0+$, hence it only remains to verify boundedness for $\delta > 1$. To this end we break up the integral in (2.5) and estimate the terms separately; let $x \in X_+$ and $\delta > 1$ then

$$\int_{-\infty}^{\infty} \frac{U(s)x}{s+i\delta}\,ds = \int_{|s|\le\delta} \frac{U(s)x}{s+i\delta}\,ds + \int_{|s|>\delta} \left(\frac{1}{s+i\delta} - \frac{1}{s}\right)U(s)x\,ds + \int_{|s|>\delta} \frac{U(s)x}{s}\,ds.$$

Let $M = \sup_s \|U(s)\| < \infty$. The first term is bounded by $2M\|x\|$. The third term is uniformly bounded by the Banach-Steinhaus theorem by $\kappa\|x\|$, for some $\kappa > 0$. The second term is bounded by

$$M\|x\| \int_{|s|>\delta} \left|\frac{\delta}{s(s+i\delta)}\right|\,ds = 2M\|x\| \int_1^\infty \frac{dr}{r\,|r+i|} \le 2M\|x\| \int_1^\infty r^{-2}\,dr = \frac{2M}{3}\|x\|,$$

which concludes the proof.

Let $X = X_0 \oplus X_+ \oplus X_-$ be the Hardy space decomposition with respect to U. As we assume A to be injective, the space X_0 is trivial. Then the restriction U_+ of U to X_+ and the restriction $U_-(-s)$ of $U(-s)$ to X_- are traces of bounded holomorphic semigroups of angle $\pi/2$. Denote the generator of the bounded holomorphic semigroup on X_+ by G_+ and the one on X_- by G_-. Then iG_+ is the restriction of A to X_+ and $-iG_-$ is the restriction of A to X_-. As is well known, generators of bounded holomorphic semigroups are just the sectorial operators of type 0. Hence the decomposition $X = X_+ \oplus X_-$ is the unique spectral decomposition corresponding to the projection associated to the lower half-plane; that is the spectral projection P corresponding to the upper half-plane is equal to $P_- = \frac{1}{2}((H^U)^2 - H^U) = \frac{1}{2}(I - H^U)$.

We summarize these considerations in the following theorem.

2.4.4 Theorem. *Let U be a bounded C_0-group with injective generator A on the Banach space X. Consider the following statements:*

1. *X is a UMD Banach space;*

2. *$H^U_{\epsilon,T}$ admits a strong limit H^U as $\epsilon \to 0$ and $T \to \infty$.*

3. *X admits a spectral decomposition with respect to A; moreover, the spectral projection is $P = \frac{1}{2}(I - H^U)$;*

4. *$H^U_{\epsilon,1}$ admits a strong limit as $\epsilon \to 0$.*

Then the following implications are true: 1. $\Rightarrow$ 2., 2. $\Rightarrow$ 3. and 3. $\Rightarrow$ 4.; if we assume A to be invertible, the assertions 2 – 4 are equivalent.

Proof. Let us consider the implication 3. $\Rightarrow$ 4. If A_1 and A_2 denote the reduced operators induced by the spectral decomposition, it is easy to see that $-iA_1$ and iA_2 are sectorial operators of type 0, which implies that $U|_{X_1}$ is the trace of a holomorphic semigroup of angle $\pi/2$, hence the corresponding analytic generator C_1 is bounded; and similarly, the analytic generator C_2 of $U|_{X_2}$ is invertible. Observe that the analytic generator C of U is given by the direct sum $C_1 \oplus C_2$. By Proposition 2.4.3 applied to the factor spaces X_1 and X_2 we have that $\mathcal{D}(C_1) + \mathcal{R}(C_1) = X_1$ and $\mathcal{D}(C_2) + \mathcal{R}(C_2) = X_2$, whence $\mathcal{D}(C) + \mathcal{R}(C) = (\mathcal{D}(C_1) \oplus \mathcal{D}(C_2)) + (\mathcal{R}(C_1) \oplus \mathcal{R}(C_2)) = (\mathcal{D}(C_1) + \mathcal{R}(C_1)) \oplus (\mathcal{D}(C_2) + \mathcal{R}(C_2)) = X_1 \oplus X_2 = X$; which implies 4. by an another application of Proposition 2.4.3. It remains to observe that, if $\mathcal{R}(A) = X$, the implication 4. $\Rightarrow$ 2. is an immediate consequence of Remark 2.4.9. □

We summarize the case of an invertible generator in the following corollary.

2.4.5 Corollary. *Let U be a bounded C_0-group on the Banach space X with invertible generator A. Then the following assertions are equivalent:*

1. $\rho(C) \neq \emptyset$, where C denotes the analytic generator of U;

2. $H^U_{\epsilon,T}$ admits a strong limit H^U as $\epsilon \to 0$, $T \to \infty$;

3. the spectral projection P is bounded; moreover $P = \frac{1}{2}(I - H^U)$.

2.4.6 Remark. If A generates a periodic group, it is known the spectrum $\sigma(A)$ of the generator consists purely of point spectrum [54, IV.2.26], and hence, if A is injective, it is automatically invertible. See also [97, Section 4.2.3] for the context of a periodic group.

2.4.7 Example. Considering the translation group on the space $L_p(\mathbb{T}, X)$ one obtains a periodic version of Example 2.3.1; it suffices to use an appropriate discrete Fourier multiplier theorem. One can always assume the operator to be injective by considering its injective part.

In [54, Ex. IV.2.30] one can find the version on the space of periodic 2π-continuous functions. Let A denote the injective part of the generator of the periodic group of translations on $C_{2\pi}(\mathbb{R})$; then the spectral projection corresponding to the upper half-plane is not bounded. Clearly, A does not act on a UMD Banach space.

2.4.2 Functional calculus approach

In the following we will recover and complement the characterization of the boundedness of the spectral projection in terms of the Hilbert transform by a direct and simpler argument via the functional calculus.

In this paragraph we will assume that the set $\mathcal{D}(A) \cap \mathcal{R}(A)$ is dense in X; recall that as A generates a (semi-)group we have $\overline{\mathcal{D}(A)} = X$. The other requirement $\overline{\mathcal{R}(A)} = X$ implies that A is injective; for reflexive spaces these two assertions are equivalent.

First we study the Hilbert transform associated with the bounded group in more detail.

2.4.8 Lemma. *For all $x \in \mathcal{D}(A) \cap \mathcal{R}(A)$ and all $d \geq 0$ the principal value integral*

$$H_d^U x := \frac{i}{\pi} PV\!\!\int_{\mathbb{R}} e^{-d|s|} \frac{U(s)x}{s}\, ds := \frac{i}{\pi} \lim_{r \to 0, R \to \infty} \int_{r \leq |s| \leq R} e^{-d|s|} \frac{U(s)x}{s}\, ds$$

exists and $H_d^U x \to H^U x := H_0^U x$ for $d \to 0+$.

Proof. Let $0 < r < 1 < R < \infty$, then

$$\int_{r \leq |s| \leq R} e^{-d|s|} \frac{U(s)x}{s}\, ds =$$

$$\int_r^1 e^{-ds} \left[\frac{(U(s) - I)x}{s} + \frac{(I - U(-s))x}{s} \right] ds + \int_1^R \frac{e^{-ds}}{s} U(s)x\, ds - \int_1^R \frac{e^{-ds}}{s} U(-s)x\, ds.$$

As $x \in \mathcal{D}(A)$, the factor in the squared brackets has a continuous extension to $[0, 1]$, hence the integral has a limit as $r \to 0$. We take care of the other two terms by integration by parts; as $x = Ay \in \mathcal{R}(A)$, the function $s \mapsto \pm U(\pm s)y$ is an anti-derivative of the function $s \mapsto U(\pm s)x$, therefore

$$\int_1^R \frac{e^{-ds}}{s} U(s)x\, ds + \int_1^R \frac{e^{-ds}}{s} U(-s)x\, ds =$$

$$\left[\frac{e^{-ds}}{s} U(s)y \right]_{s=1}^R + \int_1^R \left(\frac{dse^{-ds} + e^{-ds}}{s^2} \right) U(s)y\, ds$$

$$- \left[-\frac{e^{-ds}}{s} U(-s)y \right]_{s=1}^R + \int_1^R \left(\frac{dse^{-ds} + e^{-ds}}{s^2} \right) U(-s)y\, ds.$$

As $\sup_{t>0} te^{-t} = e^{-1} < \infty$, we deduce that also the limit as $R \to \infty$ exists. We have proved the existence of $H_d^U x$ for all $x \in \mathcal{D}(A) \cap \mathcal{R}(A)$. The fact that $H_d^U x$ converges to $H_0^U x$ as $d \to 0+$ is a consequence of Lebesgue's theorem.   □

2.4.9 Remark. We note for reference. The proof of the foregoing lemma shows in particular that the limit

$$\lim_{r \to 0} \int_{r \leq |s| \leq 1} \frac{U(s)x}{s}\, ds$$

exists for all $x \in \mathcal{D}(A)$ and that

$$\lim_{R \to \infty} \int_{1 \leq |s| \leq R} \frac{U(s)x}{s}\, ds$$

exists for all $x \in \mathcal{R}(A)$.

We will say that H^U is bounded if the densely defined operator H^U has a bounded extension to all of X. This is the case if, for example, the principal value integral exists for all $x \in X$. By the theorem above H^U is bounded if X is a UMD space.

Next we will relate the Hilbert transform with the resolvent of A by means of the Fourier transform. Observe that for $x \in \mathcal{D}(A)$ we have the inequality

$$\|(\pm d + is)R(\pm d + is, A)x\| = \|x + R(\pm d + is, A)Ax\| \leq \|x\| + Cd^{-1}\|Ax\|.$$

By iteration and using the resolvent equation this implies that, for $x \in \mathcal{D}(A^2)$, the function $s \mapsto \|R(d+is, A)x - R(-d+is, A)x\|$ is majorized by $s \mapsto Cd^{-1}|d+is|^{-2}$, with $C = C(x) \geq 0$, which is integrable on $\mathbb{R}$.

2.4.10 Lemma. Let $d > 0$ and $x \in \mathcal{D}(A^2)$, then

$$H_d^U x = \frac{1}{2\pi} \int_{\mathbb{R}} \operatorname{sign} s \, (R(-d+is, A)x - R(d+is, A)x) \, ds. \tag{2.6}$$

Proof. We have $\int_{r \leq |s| \leq R} e^{-d|s|}\frac{U(s)x}{s}\,ds = \int_{-\infty}^{\infty} h_{r,R}(s)e^{-d|s|}U(s)x\,ds$, where $h_{r,R}$ denotes the function $s \mapsto \frac{1}{s}1_{\{r \leq |u| \leq R\}}(s) \in L_p(\mathbb{R})$, for all $p \in [1, \infty]$. As the resolvent is given by the Laplace transform of its semigroup, we obtain the relation

$$\int_{\mathbb{R}} e^{-ist}e^{-ds}1_{(0,\infty)}(s)U(s)\,ds = R(d+it, A);$$

considering the reflected semigroup $(U(-s))$ generated by $-A$ this translates to

$$\int_{\mathbb{R}} e^{-ist}e^{-ds}1_{(0,\infty)}(s)U(-s)\,ds = R(d+it, -A).$$

Substituting s by $-s$ in the second integral we obtain the identity

$$\mathcal{F}(e^{-d|\cdot|}U(\cdot))(t) = \int_{\mathbb{R}} e^{-ist}e^{-d|s|}(s)U(s)\,ds = R(d+it, A) - R(-d+it, A). \tag{2.7}$$

Recall that given two functions $f \in L_1(\mathbb{R})$, $g \in L_1(\mathbb{R}, Y)$ for some Banach space Y, we have as a consequence of Fubini's theorem the relation $\int_{\mathbb{R}} \check{f}(t)g(t)\,dt = \int_{\mathbb{R}} f(t)\check{g}(t)\,dt$, where $\check{f}$ denotes the inverse Fourier transform of f. By the inversion formula, applying this relation to the identity (2.7) we find

$$\int_{r \leq |s| \leq R} e^{-d|s|}\frac{U(s)x}{s}\,ds = \int_{-\infty}^{\infty} h_{r,R}(s)e^{-d|s|}U(s)x\,ds$$

$$= \int_{-\infty}^{\infty} \check{h}_{r,R}(s)((R(d+is, A)x - R(-d+is, A)x))\,ds.$$

We compute the inverse Fourier transform of $h_{r,R}$: for $s \in \mathbb{R} \setminus \{0\}$ we have

$$\check{h}_{r,R}(s) = (2\pi)^{-1}\int_{r}^{R} t^{-1}(e^{ist} - e^{-ist})\,dt = i\pi^{-1}\int_{r/s}^{R/s} \frac{\sin u}{u}\,du;$$

it is well known from calculus that the improper integral $\int_0^{\infty} u^{-1}\sin u\,du$ converges and has as value $\pi/2$. Therefore, the functions $\check{h}_{r,R}$ are uniformly bounded on $\mathbb{R}$ and converge pointwise (as $r \to 0$ and $R \to \infty$) to the limit $i2^{-1}\mathrm{sign}\,(\cdot)$. Hence, Lebesgue's theorem implies

$$H_d^U x = \frac{1}{2\pi} \int_{\mathbb{R}} \mathrm{sign}\, s \left(R(-d + is, A)x - R(d + is, A)x \right) ds,$$

which proves the claim.   □

Using this representation, we will compare the Hilbert transform H^U and the spectral projection P. Recall that we suppose the generator A of the bounded group U to have dense range.

Denote by Σ the bisector $\Sigma_{\alpha,\omega}$ with $\alpha = (-\pi/2, \pi/2)$ and $\omega = (\pi/3, \pi/3)$; further denote by Γ the contour $\partial \Sigma_{\alpha,\tilde{\omega}}$ with positive orientation and $0 < \tilde{\omega} < \omega$.

For $d > 0$ we denote by Γ_d the contour obtained by deforming Γ as indicated in the Figure 2.1.

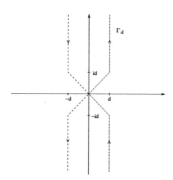

Figure 2.1: The contour Γ_d.

Denote by $p \in H^\infty(\mathbb{C} \setminus \mathbb{R})$ the indicator function of the upper complex half-plane and define the signum-function s by $s(z) = p(z) - p(-z) = 2p(z) - 1$; then $p = (1 + s)/2$. Thus, the projection $P = p(A)$ is bounded if and only if the operator $s(A)$ is bounded, where both operators are defined by means of the $H^\infty(\Sigma)$-functional calculus. We choose the following approximations of the identity $\varphi_n(z) = e^{is(z)z/n}$; the functions φ_n are in $H^\infty(\Sigma)$ with norms uniformly bounded by 1 and converge pointwise to the function which is identically 1. The operator $s(A)$ is approximated by the bounded operators $S_n = (s\varphi_n)(A)$. Indeed, for $x \in \mathcal{D}(A) \cap \mathcal{R}(A) \subset \mathcal{D}(P) = \mathcal{D}(s(A))$ we have

$$s(A)x = \lim_{n \to \infty} S_n x \tag{2.8}$$

by the Convergence Lemma. The vector $S_n x$ is given by the Cauchy integral

$$S_n x = \frac{1}{2\pi i} \int_\Gamma s(z)\varphi_n(z)R(z, A)x \, dz.$$

As φ_n decays sufficiently fast at infinity, we may deform the contour Γ into the contour Γ_d without changing the value of the integral:

$$S_n x = \frac{1}{2\pi i} \int_{\Gamma_d} s(z) \varphi_n(z) R(z, A) x \, dz \qquad \forall d > 0.$$

For $d > 0$ we define also the contour $\tilde{\Gamma}_d$ as $\{w \in \mathbb{C} : |\mathcal{R}e(w)| = d\}$ with positive orientation, and we consider the operators

$$Q_{d,n} x = \frac{1}{2\pi i} \int_{\tilde{\Gamma}_d} s(z) \varphi_n(z) R(z, A) x \, dz.$$

Let $x \in \mathcal{D}(A^2) \cap \mathcal{R}(A)$. Observe that, if $x = Ay \in \mathcal{R}(A)$, we have the identity $R(z, A)x = -y + z R(z, A)y$. Consequently, the function $R(\cdot, A)x$ is bounded on the complement of $\overline{\Sigma_{\alpha, \nu}}$ for all $\nu > 0$. The integrand being a bounded function (uniformly in n), we deduce that there is a constant $C'_x > 0$, independent of n and d, such that

$$\|Q_{d,n} x - S_n x\| \le C'_x d. \tag{2.9}$$

As n tends to infinity, $S_n x$ converges to $s(A)x = (2P - I)x$. Hence, it remains to study the behaviour of the expression $Q_{d,n} x$:

$$Q_{d,n} x = (2\pi)^{-1} \int_{-\infty}^{\infty} s(is) \varphi_n(d + is) R(d + is, A)x - s(-is) \varphi_n(-d - is) R(-d - is, A)x \, ds.$$

As $s(-is) = \text{sign}(-s) = -s(is)$ and as $\varphi_n(-z) = \varphi_n(z)$, we obtain

$$Q_{d,n} x = (2\pi)^{-1} \int_{-\infty}^{\infty} s(is) \varphi_n(d + is) \big(R(d + is, A)x + R(-d - is, A)x \big) \, ds$$

$$= (2\pi)^{-1} \int_{-\infty}^{\infty} s(is) \varphi_n(d + is) \big[R(d + is, A)x - R(-d + is, A)x \big] \, ds \tag{2.10}$$

$$+ (2\pi)^{-1} \int_{-\infty}^{\infty} s(is) \varphi_n(d + is) \big[R(-d + is, A)x + R(-d - is, A)x \big] \, ds.$$

We consider the first integral in (2.10). As $x \in \mathcal{D}(A^2)$, the function in the parentheses is integrable; hence we may apply Lebesgue's theorem, which tells us that the first term converges to $-H_d^U x$ as n tends to infinity. In order to find an estimate for the second integral in (2.10) note that by substituting s by $-s$ we have

$$(2\pi)^{-1} \int_{-\infty}^{\infty} s(is) \varphi_n(d + is) \big[R(-d + is, A)x + R(-d - is, A)x \big] \, ds$$

$$= (2\pi)^{-1} \int_{-\infty}^{\infty} s(is) \big[\varphi_n(d + is) R(-d + is, A)x - \varphi_n(d - is) R(-d + is, A)x \big] \, ds$$

$$= (2\pi)^{-1} \int_{-\infty}^{\infty} s(is) \big[\varphi_n(d + is) - \varphi_n(-d + is) \big] R(-d + is, A)x \, ds.$$

Rewriting the term in the squared brackets

$$\varphi_n(d + is) - \varphi_n(-d + is) = \varphi_n(-d + is)\left(e^{2is(is)d/n} - 1\right)$$

the second integral in (2.10) can be rewritten as

$$-(e^{-2id/n} - 1)(2\pi)^{-1} \int_{-\infty}^{0} \varphi_n(-d + is)R(-d + is, A)x\, ds$$

$$+ (e^{2id/n} - 1)(2\pi)^{-1} \int_{0}^{\infty} \varphi_n(-d + is)R(-d + is, A)x\, ds.$$

By Cauchy's theorem we may deform the path of integration $i(-\infty, 0) - d$ into the path $\gamma = (-1 - i)(0, \infty) - d$ without changing the value of the integral, and similarly we may deform the path $i(0, \infty) - d$ into $(-1 + i)(0, \infty) - d$. Thus, we obtain by the sectoriality of A and as $x \in \mathcal{R}(A)$ a constant C_x''' independent of n and d such that

$$\left\| \int_{-\infty}^{0} \varphi_n(-d + is)R(-d + is, A)x\, ds \right\| = \left\| \int_{\gamma} \varphi_n(z)R(z, A)x\, dz \right\| \qquad (2.11)$$

$$\leq C_x''' \int_{\gamma} |\varphi_n(z)|\, |dz| = C_x'''n.$$

The same estimate also holds for the second term. Hence, the second integral in (2.10) is estimated by

$$\left(\left| (e^{2id/n} - 1) \right| + \left| (e^{-2id/n} - 1) \right| \right)(2\pi)^{-1}C_x'''n = 2\sin(d/n)\pi^{-1}C_x'''n, \qquad (2.12)$$

which implies

$$\limsup_{n \to \infty} \left\| Q_{d,n}x + H_d^U x \right\| \leq 2\pi^{-1}C_x'''d. \qquad (2.13)$$

Putting all this together, we find for $x \in \mathcal{D}(A^2) \cap \mathcal{R}(A)$ and for all $d > 0$ and $n \in \mathbb{N}$

$$\left\| s(A)x + H^U x \right\| \leq \left\| s(A)x - S_n x \right\| + \left\| S_n x - Q_{d,n}x \right\| + \left\| Q_{d,n}x + H_d^U x \right\| + \left\| H_d^U x - H^U x \right\|;$$

taking the limes superior with respect to n, we obtain by (2.8), (2.9) and (2.13)

$$\left\| s(A)x + H^U x \right\| \leq C_x'd + 2C_x'''\pi^{-1}d + \left\| H_d^U x - H^U x \right\|,$$

from which we conclude applying Lemma 2.4.8, by letting $d \to 0$, that

$$\left\| s(A)x + H^U x \right\| = 0.$$

Hence, the operators H^U and $I - 2P$ agree on the dense set $\mathcal{D}(A^2) \cap \mathcal{R}(A)$. We summarize these considerations in the following theorem.

2.4.11 Theorem. *Let A be the generator of the bounded C_0-group U. Assume that A has dense range. Then the following assertions are equivalent:*

1. *The spectral projection $P = p(A)$ is bounded;*

2. *The Hilbert transform H^U has a bounded extension to X; more precisely $H^U = I - 2P$.*

2.5   The general case

In this section we consider a multisectorial operator $A \in \mathrm{Sect_d}(\alpha, \omega)$ with $\alpha \in (-\pi, \pi]^N$ and $\omega \in [0, \pi)^N$ admissible. An iteration of the steps considered in the bisectorial case gives immediately the following. There is at most one decomposition $X = X_1 \oplus \cdots \oplus X_N$ of X into a direct sum of A-invariant subspaces X_i such that the restrictions $A_i = A|_{X_i}$ are in $\mathrm{Sect_d}(\alpha_i, \omega_i)$. This decomposition exists if and only if all the projections P_i obtained by inserting the indicator functions of the sets $\Sigma_{\alpha_i, \omega_i'}$ into the functional calculus are bounded. In this case we have $X_i = \mathcal{R}(P_i)$, all the projections P_i commute and they sum up to I.

In the following we study the spectral decomposition of an (almost) R-sectorial operator. Applying Proposition 1.6.6 to the reduced operators A_i with $N(z) = z(z + A_i)^{-1}$ or $N(z) = zA_i(z + A_i)^{-2}$ gives the following Corollary.

2.5.1 Corollary. *Let $A \in \mathrm{Sect_d}(\alpha, \omega)$. Assume that $A \in \mathrm{RSect}(\alpha, \tilde{\omega})$ ($A \in \mathrm{aRSect}(\alpha, \tilde{\omega})$ resp.) with $\alpha \in (-\pi, \pi]^N$ and $\omega \leq \tilde{\omega} \in [0, \pi)^N$ admissible. Assume that the spectral projections are bounded and denote the reduced operators by A_i.*
      Then $A_i \in \mathrm{Sect_d}(\alpha_i, \omega_i) \cap \mathrm{RSect}(\alpha_i, \tilde{\omega}_i)$ ($A_i \in \mathrm{Sect_d}(\alpha_i, \omega_i) \cap \mathrm{aRSect}(\alpha_i, \tilde{\omega}_i)$ resp.).

Proof. The boundedness of the spectral projections implies that $A_i \in \mathrm{Sect_d}(\alpha_i, \omega_i)$. Consequently, the two functions N are bounded. Hence, an application of Proposition 1.6.6 completes the proof. □

2.5.2 Remark. Let $A \in \mathrm{Sect_d}(\alpha, \omega)$ have a bounded spectral decomposition. Choose complex numbers $w_1, \ldots, w_N$ that sum up to zero. Then the same reasoning that led up to Theorem 2.2.9 tells us that the operator $(e^{i(\pi-\eta_1)}A)^{w_1} \ldots (e^{i(\pi-\eta_N)}A)^{w_N}$ is a linear combination of the spectral projections P_i.

Completion of Theorem 1.10.12   The following result is a version of Proposition 1.10.7 in the setting of multisectorial operators.

2.5.3 Proposition. *Let $A \in \mathrm{aRSect_d}(\alpha, \omega)$ and $\omega' > \omega$. For every $\psi \in H_0^\infty(\Sigma_{\alpha, \omega'})$ there is a $\varphi \in H_0^\infty(\Sigma_{\alpha, \omega'})$ and constants $\gamma_1, \ldots, \gamma_N$ such that for an admissible contour Γ and $t > 0$*

$$\psi(tA) = \frac{1}{2\pi i} \int_\Gamma \frac{\varphi(\lambda)}{\lambda}[tA\lambda R(\lambda, tA)^2]d\lambda + \left(\sum_{k=1}^N \gamma_k P_k\right) tA(1 + tA)^{-2},$$

where P_k denotes the spectral projection associated to the k-th component of $\Sigma_{\alpha, \omega'}$. In particular, if all the spectral projections P_k are bounded, the set $\{\psi(tA) : t > 0\}$ is R-bounded.

Proof. The proof follows along the lines of the one for Proposition 1.10.7; with the necessary modification observed in Remark 1.10.8. □

Let us take a second look at Theorem 1.10.12. As all the square function norms are equivalent by Proposition 1.10.4, we deduce from Theorem 1.10.12 and Lemma 1.10.6 the following result. We will use the notation introduced in the beginning of Section 1.10.

2.5.4 Corollary. *Let* $A \in \mathrm{Sect}_d(\alpha, \omega) \cap \mathrm{aRSect}(\alpha, \tilde{\omega})$, $\omega \leq \tilde{\omega}$ *admissible. Assume that we have also the equivalence of norms* (1.11).

Then, for all $k \in \{1, \ldots, N\}$, *the operator* $A_k = e^{i(\pi - \eta_k)} A$ *has a bounded sectorial* H^∞*-functional calculus.*

In particular, all the associated spectral projections are bounded by Theorem 2.2.9. Consequently, we obtain by Proposition 2.5.3 (the proof is analogue to the proof given for Lemma 1.10.6)

2.5.5 Proposition. *Assume the equivalence* (1.11) *of the square function norms to the norm in* X. *If* $A \in \mathrm{aRSect}_d(\alpha, \omega)$ *and* $\omega < \tilde{\omega}$, *then the set* $\{z^{1/2} A^{1/2} R(z, A) : z \in \mathbb{C} \backslash \overline{\Sigma_{\alpha, \tilde{\omega}}}\}$ *is R-bounded.*

Thus, we obtain the following characterization of the boundedness of the multisectorial functional calculus.

2.5.6 Theorem. *Let* $A \in \mathrm{Sect}_d(\alpha, \omega) \cap \mathrm{aRSect}(\alpha, \tilde{\omega})$, $\omega \leq \tilde{\omega}$ *admissible. Then* A *has a bounded functional calculus if and only if for one (all) square functions we have the equivalence* (1.11). *In this case we have* $\omega_{H^\infty}(A) = \omega_r(A)$.

Proof. Combine Proposition 2.5.5 with Theorem 1.10.12. For the angles, observe that, if A has a bounded H^∞-functional calculus, we obtain by 2. of Theorem 1.10.12 the relation $\omega_{H^\infty}(A) \leq \tilde{\omega}$; taking the infimum yields $\omega_{H^\infty}(A) \leq \omega_r(A)$. The reverse inequality holds by Corollary 1.10.10. $\square$

If A has a bounded multisectorial functional calculus, all the reduced operators have bounded sectorial functional calculi. On the other hand, we may construct a bounded multisectorial functional calculus, if we know that all the operators $A_k = e^{i(\pi - \eta_k)} A$ admit a bounded sectorial functional calculus. Indeed, then the spectral projections are bounded. Denoting the reduced operators by B_k we find by Corollary 2.5.1 that all operators B_k are almost R-sectorial. Furthermore, we may reduce the domain of holomorphy of the functional calculus for B_k by the optimality of angles. Now, we are in a situation that the functional calculi for $B_1, \ldots, B_N$ are defined on disjoint sectors. Considering the direct sum, we may build a bounded multisectorial functional calculus for the operator A.

Of course, this and more is contained already in Theorem 1.10.12 and the remarks following it. However, this more constructive approach has the advantage of being simpler (assuming that the case of a sectorial operator is already understood), as it does not make again appeal to the sophisticated (but powerful) tool of square functions.

2.6   Applications of the spectral decomposition

In this section we will demonstrate how the boundedness of the spectral projections allows to transfer results from the sectorial setting to the multisectorial setting. First we will consider a result on the joint functional calculus due to Lancien, Lancien and Le Merdy. Let us recall their result (see [83] or [74, Cor. 6.2] and compare with Theorem 1.7.11).

2.6.1 Theorem. *Let X be a Banach space with property (A) and let $A \in \mathrm{Sect}_\mathrm{d}(\omega_A)$, $B \in \mathrm{Sect}_\mathrm{d}(\omega_B)$ be two commuting operators that have a bounded $H^\infty(\Sigma_{\sigma_A})$ resp. a bounded $H^\infty(\Sigma_{\sigma_B})$-functional calculus. Then for any $\sigma > \sigma_A$ and $\sigma' > \sigma_B$ the pair (A, B) has a bounded $H^\infty(\Sigma_\sigma \times \Sigma_{\sigma'})$-functional calculus.*

Property (A) is weaker than property (α). For its definition and further background information we refer to [83]. Applying this result we obtain as a corollary.

2.6.2 Corollary. *Let X be a Banach space with property (A) and let $A \in \mathrm{Sect}_\mathrm{d}(\alpha, \omega_A)$, $B \in \mathrm{Sect}_\mathrm{d}(\beta, \omega_B)$ be two commuting operators that have a bounded $H^\infty(\Sigma_{\alpha,\omega'_A})$ resp. $H^\infty(\Sigma_{\beta,\omega'_B})$- functional calculus. Then the pair (A, B) has a bounded $H^\infty(\Sigma_{\alpha,\omega} \times \Sigma_{\beta,\omega'})$- functional calculus for all $\omega > \omega'_A$, $\omega' > \omega'_B$.*

Proof. The idea is to decompose the Banach space in order to apply Thm. 2.6.1 to the reduced operators. Since A has a bounded $H^\infty(\Sigma_{\alpha,\omega'_A})$-functional calculus, the space X decomposes into a direct sum $X = X_1 \oplus \cdots \oplus X_{N_A}$ where the X_i are A-invariant subspaces such that the restrictions $A_i = A|_{X_i}$ are in $\mathrm{Sect}_\mathrm{d}(\alpha_i, \omega_{A,i})$. Clearly, each A_i has a bounded $H^\infty(\Sigma_{\alpha_i,\omega'_{A,i}})$-functional calculus. Since A and B commute, the X_i are also B-invariant and we denote the restriction of B to X_i by B_i. As each B_i has a bounded $H^\infty(\Sigma_{\beta,\omega'_B})$-functional calculus, we can in turn decompose each X_i in a direct sum $X_i = X_i^1 \oplus \cdots \oplus X_{N_B}^i$ of subspaces with analogous properties for B_i. As before each X_j^i is also A_i-invariant.

In this way we obtain on each of the spaces X_j^i a pair $(A_{i,j}, B_{i,j})$ of commuting operators satisfying the assumptions of Thm. 2.6.1; we remark that Thm. 1.4.24 allows to rotate all operators suitably. Therefore, by Thm. 2.6.1, on each space X_j^i the pair $(A_{i,j}, B_{i,j})$ has a bounded $H^\infty(\Sigma_{\alpha_i,\omega_i} \times \Sigma_{\beta_j,\omega'_j})$-functional calculus. As the pair (A, B) is the direct sum of these, it is easy to see that (A, B) has a bounded $H^\infty(\Sigma_{\alpha,\omega} \times \Sigma_{\beta,\omega'})$- functional calculus. $\square$

Now, we will use the spectral decomposition to transfer a result from [39, 74] on the invertibility of a sum of sectorial operators to the multisectorial setting. This result complements the closedness result of Theorem 1.7.7. In particular, it shows that the range of $A + B$ is dense in X. Moreover, it removes the assumption that B has dense range.

2.6.3 Theorem. *Let A and B be two densely defined resolvent commuting multisectorial operators in the Banach space X. Let $\Sigma = \Sigma_{\alpha,\omega}$ and $\tilde{\Sigma} = \Sigma_{\beta,\tilde{\omega}}$. Assume that*

1. A has a bounded $H^\infty(\Sigma)$-calculus and dense range;

2. $-B \in \mathrm{RSect}(\beta, \tilde{\omega})$;

3. the closures of the sets Σ and $\tilde{\Sigma}$ meet only in zero.

Then the operator $A + B$ with domain $\mathcal{D}(A) \cap \mathcal{D}(B)$ is closed and

$$\|Ax\| + \|Bx\| \le C \|Ax + Bx\| \tag{2.14}$$

for all $x \in \mathcal{D}(A) \cap \mathcal{D}(B)$. Moreover, $A + B$ is invertible if either A or B is invertible.

Proof. Since A has a bounded $H^\infty(\Sigma)$-calculus, we may decompose the Banach space into a direct sum $X = X_1 \oplus \cdots \oplus X_N$ where the X_i are A-invariant subspaces such that the restrictions $A_i = A|_{X_i}$ are in $\mathrm{Sect}_d(\alpha_i, \omega_{A,i})$. As A and B commute, the subspaces X_i are also invariant with respect to B. We denote the restriction of B to X_i by B_i. It suffices to show that the reduced operators $A_i + B_i$ satisfy the inequality (2.14) and are invertible on X_i. Fix $i_0 \in \{1, \ldots, N\}$. We regard A_{i_0} and B_{i_0} as sectorial operators, i.e. the spectra are contained in a single sector. For $\epsilon > 0$ there is a number $-w_\epsilon \in \tilde{\Sigma}$ of modulus ϵ such that the sectorial operator $B_{i_0} + w_\epsilon$ is invertible. Hence, combining the proof of Theorem 1.7.7 with Kahane's contraction principle we find that there is a constant C independent of $\epsilon > 0$ such that

$$\|Ax\| + \|(B + w_\epsilon)x\| \le C \|Ax + Bx + w_\epsilon x\|$$

for all $x \in \mathcal{D}(A_{i_0}) \cap \mathcal{D}(B_{i_0})$. Letting ϵ tend to zero proves the desired inequality.

The invertibility of $A_{i_0} + B_{i_0}$ follows directly from [39, Théorème 3.7] by a quite similar argument. $\square$

Norm estimates for Riesz transforms.   Next we will extend Example 2.3.2 to N dimensions, i.e. we want to establish the relation

$$\left\|(-\Delta)^{1/2} f\right\| \sim \sum_{i=1}^N \left\|\frac{d}{dx_i} f\right\| \qquad \forall f \in \mathcal{D}((-\Delta)^{1/2}) \cap W^{1,p}(\mathbb{R}^N, X).$$

We will use the spectral decomposition instead of a randomization argument as in Theorem 2.3.4. The operator $(-\Delta)^{1/2}$ is the generator of the Poisson semigroup.

2.6.4 Example. Let X be a UMD Banach space with property (α) and $1 < p < \infty$. Then the UMD space $L_p(\mathbb{R}^N, X)$ has property (α) [81, 4.10]. Let $A_i = \frac{d}{dx_i}$ be the generator of the translation group acting on the ith component. As in Example 2.3.1 it follows that A_i is a bisectorial operator, more precisely $A_i \in \mathrm{Sect}_d(\alpha, \omega)$ with $\alpha = (-\pi/2, \pi/2)$ and $\omega = 0$. The operators A_i commute (as the translation groups clearly do so) and possess a bounded $H^\infty(\Sigma_{\alpha, \omega'})$-functional calculus for each admissible $\omega' > 0$. Hence the claim follows from the following theorem.

The same argument as in the one-dimensional case (as $\mathcal{D}(-\Delta) = W^{2,p}(\mathbb{R}^N, X)$ is dense in the Sobolev space $W^{1,p}(\mathbb{R}^N, X)$ and a core for $(-\Delta)^{1/2}$) allows also in this case to determine the domain of $(-\Delta)^{1/2}$; it is given by the Sobolev space $W^{1,p}(\mathbb{R}^N, X)$.

2.6.5 Theorem. *Let X be a Banach space with property (α) and $N \in \mathbb{N}$. For $1 \le i \le N$ let $A_i \in \mathrm{Sect}_d(\alpha, \omega)$ with $\alpha = (-\pi/2, \pi/2)$ and $\omega \in [0, \pi/4)^2$. Assume that the operators A_i are densely defined with dense range, commute and have a bounded $H^\infty(\Sigma_{\alpha, \omega'})$-functional calculus for some $\omega \le \omega' \in [0, \pi/4)^2$. Then we have*

$$\left\|(-(A_1^2 + \cdots + A_N^2))^{1/2} f\right\| \sim \sum_{i=1}^N \|A_i f\|$$

for all $f \in \mathcal{D} := \mathcal{D}((-(A_1^2 + \cdots + A_N^2))^{1/2}) \cap \bigcap_{i=1}^N \mathcal{D}(A_i)$.

Proof. We denote by P_i the bounded spectral projection 'onto the upper half-plane' associated to A_i. By Theorem 1.7.11 the N-tuple $\mathbf{A} = (A_1, \dots, A_N)$ has a bounded $H^\infty(\Sigma_{\alpha,\omega'}^N)$-functional calculus. Observe that $-(A_1^2 + \cdots + A_N^2)$ is a sectorial operator. The functions $h_i : z \mapsto (-(z_1^2 + \cdots + z_N^2))^{-1/2} z_i$ are in $H^\infty(\Sigma_{\alpha,\omega'}^N)$, $1 \le i \le N$, whence we have that

$$c \, \|A_i f\| \le \left\| (-(A_1^2 + \cdots + A_N^2))^{1/2} f \right\|$$

for all $f \in \mathcal{D}$ and some constant $c > 0$. In particularly, we find

$$cN^{-1} \sum_{i=1}^N \|A_i f\| \le \left\| (-(A_1^2 + \cdots + A_N^2))^{1/2} f \right\|.$$

In order to obtain the reverse inequality we would like to plug $\mathbf{A}$ into the function $h(z) = (-(z_1^2 + \cdots + z_N^2))^{1/2} (z_1 + \cdots + z_N)^{-1}$, however h is not bounded on $\Sigma_{\alpha,\omega'}^N$. The reason being that the variables z_i may cancel out due to different 'signs'. In the following we will take care of this fact. As the spectral projections are bounded and commute with all resolvents of A_i, they induce a decomposition of X into a direct sum of 2^N subspaces X_j which are A_i-invariant. As X has property (α), so do the closed subspaces X_j. Moreover, the reduced operators $A_{i,j} = A_i|_{X_j}$ are in $\mathrm{Sect}_{\mathrm{d}}(\alpha_j, \omega_j)$ where $\alpha_j \in \{-\pi/2, \pi/2\}$ and $\omega_j \in [0, \pi/4)$. Therefore, there is a sequence of signs $\epsilon_{i,j} \in \{-1, 1\}$ such that the operators $B_{i,j} = \epsilon_{i,j} A_{i,j}$ are in $\mathrm{Sect}_{\mathrm{d}}(\pi/2, \theta)$ for some $\theta \in [0, \pi/4)$ and have for some $\theta' \in (\theta, \pi/4)$ a bounded $H^\infty(\Sigma_{\pi/2,\theta'})$-functional calculus. As the function h is bounded on $\Sigma_{\pi/2,\theta'}^N$, we obtain by the boundedness of the joint functional calculus of the vector $(B_{1,j}, \dots, B_{N,j})$ on X_j and Theorem 1.4.24 the inequality

$$\left\| (-(A_{1,j}^2 + \cdots + A_{N,j}^2))^{1/2} f \right\| = \left\| (-(B_{1,j}^2 + \cdots + B_{N,j}^2))^{1/2} f \right\|$$

$$\le C \, \|B_{1,j} f + \cdots + B_{N,j} f\| \le C \sum_{i=1}^N \|A_{i,j} f\|,$$

for $f \in \mathcal{D} \cap X_j$. Having established the desired norm estimate on each of the factor spaces X_j, we obtain the inequality on $X = X_1 \oplus \cdots \oplus X_{2^N}$ by (2.1). $\qquad\square$

Chapter 3

Maximal regularity

In the first two sections of this chapter we will study the problem of maximal L_p regularity for first and second order Cauchy problems on the real line. The property of maximal regularity is important as its regularizing effect allows to employ iteration procedures and in particular fixed point arguments. We will illustrate this in the third section where we employ maximal regularity results in order to derive the existence and uniqueness of solutions of abstract quasilinear evolution equations on the line. In that section we also study various concrete evolution equations of first and second order, for example, we consider several elliptic quasilinear equations on cylindrical domains. Another application of maximal regularity yields the existence of center manifolds of quasilinear equations; this is the content of the last section of this chapter.

We will see that the problem of maximal regularity on the real line leads in a natural way to consider bisectorial operators; in contrast to the well known case of an interval $[0, T)$, $T \geq 0$, which leads to sectorial operators.

3.1   First order problems

Let $p \in (1, \infty)$ and consider the first order Cauchy problem

$$u'(t) + Cu(t) = f(t) \qquad (t \in \mathbb{R}), \tag{3.1}$$

where C is a densely defined closed linear operator on X and $f \in L_p(\mathbb{R}, X)$.

3.1.1 Definition. We say that C (or the problem (3.1)) satisfies *maximal L_p-regularity* on the line for the Equation (3.1) if for all $f \in L_p(\mathbb{R}, X)$ there exists a unique solution $u_f \in W^{1,p}(\mathbb{R}, X) \cap L_p(\mathbb{R}, \mathcal{D}(C))$.

We may rephrase this in saying that C has maximal regularity if and only if the operator sum $A + B$ with domain $\mathcal{D}(A) \cap \mathcal{D}(B)$ is bijective, where A is the derivation operator on $W^{1,p}(\mathbb{R}, X)$ and B is defined pointwise by $(Bu)(t) = C(u(t))$ on $L_p(\mathbb{R}, \mathcal{D}(C))$. It follows from the closed graph theorem that this is equivalent to the invertibility of $A + B$; in other words, the solution operator $f \mapsto u_f$ is bounded from $L_p(\mathbb{R}, X)$ to

$W^{1,p}(\mathbb{R}, X) \cap L_p(\mathbb{R}, \mathcal{D}(A))$. The closedness of the map $f \mapsto u_f$ is a consequence of the variation of constants formula.

Mielke [92] showed that if C satisfies L_p-maximal regularity on the line then $i\mathbb{R}$ is in the resolvent set and we have the estimate

$$\|R(it, C)\| \le c(1 + |t|)^{-1} \qquad \text{for } t \in \mathbb{R}. \tag{3.2}$$

In particular, $C \in \mathrm{Sect}_d(\beta, \omega)$ for $\beta = (0, \pi)$ and some $\omega > 0$, that is, C is canonically bisectorial.

Observe that, in contrast to the case of L_p-maximal regularity on a finite interval $[0, T)$, the closedness of the operator sum $A + B$ is not equivalent to its invertibility. Indeed, let C denote the Laplacian on $L_p(\mathbb{R}^N)$ with domain $W^{2,p}(\mathbb{R}^N)$. Then C does not satisfy maximal regularity on the line, as C is not invertible. However, $-C$ admits a bounded H^∞-functional calculus of angle 0, in particular the operator is R-sectorial (see [74, Thm. 5.3]; the space L_p is a UMD space, thus it has property (Δ)). Therefore, we may conclude from Theorem 1.7.7 that the sum $A + B$ is closed.

Using the classical Mikhlin multiplier theorem Mielke could show that the bisectoriality condition (3.2) actually characterizes L_p-maximal regularity on the line if the underlying space is (isomorphic to) a Hilbert space. By means of Weis' extension of this multiplier theorem to UMD spaces Schweiker [108] proved that this condition is also sufficient in a UMD Banach space if the set of operators $\{isR(is, C) : s \in \mathbb{R}\}$ is assumed not only to be bounded but to be R-bounded (see also [122]).

We remark that the property of L_p-maximal regularity on the line is independent of $p \in (1, \infty)$ (see [108] for details); for that reason we will just speak of maximal regularity (on the line) for the most part.

We will apply first the Theorem 1.7.7 to recover this result on maximal L_p-regularity on the line. Let $A = \frac{d}{dt}$ be the generator of the translation group on $L_p(\mathbb{R}, X)$. Suppose that X is a UMD Banach space, then A has a bounded $H^\infty(\Sigma_{\alpha, \omega'})$-functional calculus, where $\alpha = (-\pi/2, \pi/2)$ and $\omega' > 0$ (see Example 1.5.11).

Assume that C is densely defined, invertible and satisfies the R-boundedness condition $C \in \mathrm{RSect}(\beta, \tilde{\omega})$ for some $\tilde{\omega} > 0$, $\beta = (0, \pi)$; note that this holds if the set $\{isR(is, C) : s \in \mathbb{R} \setminus \{0\}\}$ is R-bounded as can be deduced from the power series expansion of the resolvent about it

$$zR(z, C) = \sum_{k=0}^{\infty} z(it - z)^k R(it, C)^{k+1} \tag{3.3}$$

(compare [81, 2.21a)]). We denote by B the extension of C (by pointwise multiplication $(Bu)(t) = C(u(t))$) to $\mathcal{D}(B) = L_p(\mathbb{R}, \mathcal{D}(C))$ and we remark that this extension inherits the spectral properties of C. The operators A and B commute in the sense of resolvents. Indeed, the semigroups associated with the sectorial operators A and iB are the translation group and pointwise multiplication by e^{tiC}, which commute. As the resolvents are obtained as the Laplace transforms of these semigroups, we find that the resolvents commute also. The fact that $B \in \mathrm{RSect}_d(\beta, \tilde{\omega})$ combines with Theorem 1.7.7 to imply the closedness of $A + B$. Now, as B is invertible, we obtain moreover the invertibility of $A + B$ which implies L_p-maximal regularity of C. In [108] and [122] (for the interval $[0, T)$) this result was obtained directly by means of Weis's vector-valued Mikhlin multiplier theorem.

Similarly we can even recover L_p-maximal regularity results for the *periodic problem*

$$u'(t) + Cu(t) = f(t), \qquad u(0) = u(2\pi)$$

on $[0, 2\pi]$, where $p \in (1, \infty)$ and C is a closed, densely defined operator. Given an inhomogeneity $f \in L_{2\pi}^p(\mathbb{R}, X)$ we want to find a unique solution $u_f \in W_{2\pi}^{1,p}(\mathbb{R}, X) \cap L_{2\pi}^p(\mathbb{R}, \mathcal{D}(C))$, where $L_{2\pi}^p(\mathbb{R}, X)$ and $W_{2\pi}^{1,p}(\mathbb{R}, X)$ denote the subspaces of $L_p(\mathbb{R}, X)$ and $W^{1,p}(\mathbb{R}, X)$ respectively, of 2π-periodic functions (see [9, 10] for some background). Let A denote the generator of the translation group on $L_{2\pi}^p(\mathbb{R}, X)$. Suppose that X is a UMD Banach space then A is a bisectorial operator and has a bounded $H^\infty(\Sigma_{\alpha,\omega'})$-functional calculus, where $\alpha = (-\pi/2, \pi/2)$ and $\omega' > 0$. Suppose that $i\mathbb{R} \subset \rho(C)$ and that the set $\{isR(is, C) : s \in \mathbb{R}\}$ is R-bounded. Denote by B the operator of multiplication by C. Then, as above, $A + B$ is invertible, which implies the L_p-maximal regularity for the periodic problem. However, these assumptions on C are unnecessarily strict.

Observe that in [9] Arendt and Bu actually characterize L_p-maximal regularity for the periodic problem (given a UMD space X and $1 < p < \infty$)) using a discrete Marcinkiewicz multiplier theorem by the condition

$$i\mathbb{Z} \subset \rho(C) \quad \text{and} \quad \mathcal{R}(\{kR(ik, C) : k \in \mathbb{Z}\}) < \infty. \tag{3.4}$$

We will show that our method also yields the sufficiency of this condition. First we deduce from it, that C is in fact asymptotically R-bisectorial.

3.1.2 Lemma. *Assume that C satisfies* (3.4). *Then C is asymptotically R-bisectorial. If moreover $i\mathbb{R} \subset \rho(C)$, then C is canonically R-bisectorial.*

Let us deduce the invertibility of $A + B$. The derivation operator A is not only bisectorial, but we find (as a consequence of the periodicity of the translation group on the torus) that $\sigma(A) = i\mathbb{Z}$. As before, the operators A and C commute. The operator C being asymptotically R-bisectorial, so is the associated operator B. The condition $i\mathbb{Z} \subset \rho(C)$ guarantees that $\sigma(A) \cap \sigma(-B) = \emptyset$. Hence, Theorem 1.7.20 is applicable and implies that $A + B$ is an invertible operator, i.e. C satisfies maximal L_p-regularity for the periodic problem.

Proof of the lemma. Indeed, the power series expansion of the resolvent (3.3) about the points ik implies that there is a radius $r > 0$ such that the sets $\tau_k = \{zR(z, C) : |z - ik| \le rk\}$ are uniformly bounded. In particular, the resolvent exists and we have a sectoriality estimate for $z \in \Sigma_{\alpha,\tilde{\omega}}$ with $|z| > c_0$ for some constant $c_0 > 0$ and a suitable $\tilde{\omega} > 0$; i.e. the operator is asymptotically bisectorial.

However, some (bounded) parts of the imaginary axis may belong to the spectrum of C. Let us remark that, if we assume that $i\mathbb{R} \subset \rho(C)$, the R-boundedness of the set $\{kR(ik, C) : k \in \mathbb{Z}\}$ implies the R-boundedness of the set $\{tR(it, C) : t \in \mathbb{R}\}$ and hence the R-bisectoriality of C. Indeed, write $t \in \mathbb{R}$ in the form $t = a_t + n_t$ where $n_t \in \mathbb{Z}$ and $a_t \in [0, 1)$. By the resolvent equation we have

$$itR(it, C) = itR(in_t, C) - ia_t tR(in_t, C)R(it, C)$$

which implies readily the R-boundedness making use of Kahane's contraction principle and the fact that the set $\{R(it, C) : t \in \mathbb{R}\}$ is R-bounded; to see this, observe that $R(it, C) = \int_{-\infty}^{\infty} R(is, C)^2 1|_{(t, \infty)}\, ds$ and that $R(i \cdot, C)^2$ is integrable by our assumptions and apply [81, Cor. 2.17]. In general, a slight modification of this argument yields that C is asymptotically R-bisectorial. □

Before turning to the second order equation, we will close the gap and show that maximal regularity on the line is indeed also characterized in terms of R-boundedness. We are considering the abstract Cauchy problem

$$u' + Au = f$$

on $\mathbb{R}$. As mentioned above, a necessary condition we have to impose on A so that we can hope for having maximal regularity is that $i\mathbb{R} \subset \rho(A)$ and that the set $\{sR(is, A) : s \in \mathbb{R}\}$ is bounded; in other words A is bisectorial and invertible.

For reference we state Weis' extension of the Mikhlin multiplier theorem; we use multi-index notation.

3.1.3 Theorem. *Let X and Y be UMD-spaces and $1 < p < \infty$. Assume that $m \in C^n(\mathbb{R}^n \setminus \{0\}, \mathcal{L}(X, Y))$ and assume that*

$$\tau = \{|t|^{|\alpha|} D^\alpha m(t) : t \in \mathbb{R}^n \setminus \{0\}, \alpha \leq (1, 1, \ldots, 1)\} \tag{3.5}$$

is R-bounded. Then m induces a bounded Fourier multiplier T_m and

$$\|T_m\|_{\mathcal{L}(L_p(\mathbb{R}^n, X), L_p(\mathbb{R}^n, Y))} \leq C\mathcal{R}_p(\tau)$$

for some constant C depending only on X, Y, n and p.

Proof. See [81, Thm. 4.6] or [114]. □

First, we will state a lemma that will be a crucial ingredient in the proof of necessity of the R-boundedness assumption. It is a sort of converse of Mikhlin's multiplier theorem (in one dimension).

3.1.4 Lemma. *Let X, Y be Banach spaces and let $M \in C^1(\mathbb{R} \setminus \{0\}, \mathcal{L}(X, Y))$. Assume that the symbol M defines a bounded L_p-multiplier from $L_p(\mathbb{R}, X)$ to $L_p(\mathbb{R}, Y)$, $1 < p < \infty$. Then the set $\{M(t) : t \in \mathbb{R} \setminus \{0\}\}$ is R-bounded.*

Proof. It suffices to note that a necessary condition on the symbol in order to give rise to a bounded L_p-multiplier is the R-boundedness of the set $\{M(s) : s$ Lebesgue point of $M\}$, see [42, Prop. 3.17]. □

3.1.5 Theorem. *Let X be a UMD Banach space and A be a densely defined bisectorial, invertible operator on X such that $\{sR(is, A) : s \in \mathbb{R}\}$ is bounded. Then the following statements are equivalent:*

1. *A satisfies L_p-maximal regularity on the line for problem* (3.1);

2. *there is a constant $C < \infty$ such that*

$$\mathcal{R}(\{a2^n R(ia2^n, A) : n \in \mathbb{Z}\}) \leq C \qquad \text{for all } 1 \leq |a| \leq 2; \qquad (3.6)$$

3. *the set $\{sR(is, A) : s \in \mathbb{R}\}$ is R-bounded.*

Proof. Taking Fourier transforms on both sides of the (3.1), we find that maximal regularity on the line for A is equivalent to the fact that the function $s \mapsto isR(is, A)$ is the symbol of a bounded L_p-multiplier; see [108, Chapter 3] for details.

The implication 2. $\Rightarrow$ 1. follows immediately from Weis' extension of Mikhlin's multiplier theorem (see [122, Rem. 3.5]) which is applicable in this case. It is easy to verify that the assumption implies that the sets $\{M(\pm 2^n) : n \in \mathbb{Z}\}$ and $\{a2^n M'(a2^n) : n \in \mathbb{Z}\}$, $1 \leq |a| \leq 2$ are uniformly R-bounded for the symbol $M(t) = AR(it, A)$.

The implication 3. $\Rightarrow$ 2. is trivial.

Finally, the implication 1. $\Rightarrow$ 3. is an immediate consequence of Lemma 3.1.4. $\square$

3.1.6 Remarks. 1. Observe that the boundedness assumption on the resolvent, i.e. the boundedness of the set $\{sR(is, A) : s \in \mathbb{R}\}$, is implied by conditions 1. and 3.

2. As we assume A to be invertible, it suffices to take $n \in \mathbb{N}$ or $|s| \geq c > 0$ in the theorem above. Note that holomorphic images of compact sets are R-bounded.

That is, maximal regularity on the line is characterized by exactly the same R-boundedness condition as L_p-maximal regularity on an interval (compare [122, Thm. 4.2], [9] (for the periodic case)). However, in the first case the occurring operators are bisectorial, in the latter setting they are sectorial.

Let A be a densely defined sectorial operator of angle $\phi < \pi/2$. Recall that we say that A *satisfies L_p-maximal regularity on* $[0, T)$, $T \in (0, \infty]$, if, given an inhomogeneity $f \in L_p([0, T), X)$, there is a unique solution $u \in W^{1,p}([0, T), X) \cap L_p([0, T), \mathcal{D}(A))$ of the abstract Cauchy problem $u' + Au = f$ on $[0, T)$ satisfying the initial condition $u(0) = 0$. If one considers maximal regularity on the line, the decay of the solution at infinity compensates for the lack of an initial condition. In the following we will briefly say that A satisfies (MR) resp. $(MR)_T$ if A satisfies L_p-maximal regularity on the line resp. the interval $[0, T)$. It is well known that this property is independent of the chosen $T \in (0, \infty)$, if A is invertible it is even independent of the choice of $T \in (0, \infty]$; this property is also independent of the chosen $p \in (1, \infty)$. Moreover, if A satisfies $(MR)_T$, it is well known [45, 85] that A is the generator of a bounded analytic C_0-semigroup, whereas in order to have maximal regularity on the line we only require A to be bisectorial (and invertible); which means that, in general, the question of maximal regularity on the line is quite different from that on an interval $[0, T)$.

In the following we consider the situation that the densely defined canonically bisectorial, invertible operator A admits a spectral decomposition. Let $X = X_1 \oplus X_2$ be the induced decomposition and denote by A_k the restriction of A to the invariant subspace X_k. Denote the spectral projection onto X_1 by P. We may assume that $-A_1$ and A_2 are sectorial (of type $< \pi/2$), that is generators of bounded holomorphic C_0-semigroups on X_1 and X_2 respectively. It is easy to see that the R-boundedness condition (3.6) for A holds if and only if both A_1 and A_2 are satisfying it. We may rephrase this as follows.

3.1.7 Corollary. *Let A be an invertible canonically R-bisectorial operator that admits a spectral decomposition as described above. Then A satisfies maximal regularity on the line if and only if $-A_1$ and A_2 satisfy maximal regularity on $[0, T)$.*

In this case the solution u_f of (3.1) corresponding to the inhomogeneity $f \in L_p(\mathbb{R}, X)$ is given by

$$u_f(t) = \int_{\mathbb{R}} K(t - s) f(s) \, ds$$

where the convolution kernel $K \in L_1(\mathbb{R}, \mathcal{L}(X))$ is given by

$$K(s) = \begin{cases} -e^{-sA_1} P, & s < 0 \\ e^{-sA_2}(I - P), & s > 0. \end{cases}$$

Indeed, u_f is the unique mild solution of the abstract Cauchy problem $u' + Au = f$ (see [108, Thm. 2.3.5] for details); hence, it is the desired (strong) solution, as we assume maximal L_p-regularity.

Since sectorial operators of type less than $\pi/2$ are clearly i-bisectorial, this allows to transfer the result of Kalton and Lancien [73] characterizing Hilbert spaces by maximal regularity to the case of maximal regularity on the line.

3.1.8 Corollary. *Let X be a Banach space that has an unconditional Schauder basis. Then either X is isomorphic to a Hilbert space or there is a densely defined i-bisectorial, invertible operator A on X that does not satisfy maximal regularity on the line.*

3.1.9 Remarks. 1. Of course, the operator A is actually a sectorial operator of type $< \pi/2$ and in this case the notions of maximal regularity on the line and on the interval $[0, T)$ coincide.

2. The Haar system is an unconditional basis for the spaces $L_p(0, 1)$, $1 < p < \infty$.

3.2   Second order elliptic equations

In the following we will consider the second order Cauchy problem

$$u''(t) + Cu(t) = f(t) \qquad t \in \mathbb{R}, \tag{3.7}$$

where C is assumed to be a closed densely defined operator on a Banach space X.

As for the first order Cauchy problem, we will say that C (or the problem (3.7)) satisfies L_p-*maximal regularity on the line*, if $p \in (1, \infty)$ and for all $f \in L_p(\mathbb{R}, X)$ there is a unique solution $u \in W^{2,p}(\mathbb{R}, X) \cap L_p(\mathbb{R}, \mathcal{D}(C))$. By the closed graph theorem, this is equivalent to the invertibility of $A + B$ on $\mathcal{D}(A) \cap \mathcal{D}(B)$, where $A = \frac{d^2}{dt^2}$ is derivation on $W^{2,p}(\mathbb{R}, X)$ and B is defined pointwise by $(Bu)(t) = C(u(t))$ on $L_p(\mathbb{R}, \mathcal{D}(C))$.

A necessary condition for maximal regularity is that $-C$ is sectorial of type $\omega \in [0, \pi)$ and invertible (see [108]). Moreover, the solution operator $f \mapsto u$ is bounded from $L_p(\mathbb{R}, X)$ to $W^{1,p}(\mathbb{R}, X) \cap L_p(\mathbb{R}, \mathcal{D}(C))$ in this case.

If the underlying Banach space has UMD, it is well known that the second derivative $-A$ on $L_p(\mathbb{R}, X)$, that is, the negative Laplacian, is sectorial of angle 0; and that $-A$ has a bounded H^∞-functional calculus of angle 0 (see Theorem 3.4.3). Hence, if C is a densely defined invertible operator on a UMD-space X with $[0, \infty) \subset \rho(C)$ and $\mathcal{R}\{\lambda R(\lambda, C) : \lambda \geq 0\} < \infty$, then it is immediate from Theorem 1.7.7 that $A + B$ is invertible. In fact, R-sectoriality is not only sufficient but characterizes maximal regularity as we will see in the next section. The argument given there is based on the Mikhlin-Weis multiplier theorem. Later we will deduce maximal regularity of the second order problem by reducing it to a first order problem on an appropriate product space.

3.2.1 Via multiplier theorem

We observe that in analogy to the first order case, considering the solution operator as a Fourier multiplier, the next theorem is a consequence of Weis' extension of the Mikhlin multiplier theorem to operator valued functions; see [108].

3.2.1 Theorem. *Let X be a UMD Banach space and $-A$ be a densely defined sectorial, invertible operator on X. Then the following statements are equivalent:*

1. *A satisfies L_p-maximal regularity on the line for problem (3.7);*

2. *the set $\{\lambda R(\lambda, A) : \lambda \in [0, \infty)\}$ is R-bounded.*

Proof. Taking Fourier transforms on both sides of equation (3.7), we find that maximal regularity on the line for A is equivalent to the fact that the function $s \mapsto s^2 R(s^2, A)$ is the symbol of a bounded L_p-multiplier; see [108, Thm. 6.2.10]. Now, the argument may be completed as in the proof of Theorem 3.1.5. $\qquad\square$

3.2.2 Via reduction to first order system

Now, we will rewrite the second order problem as a first order problem on an appropriate product space. First, we will provide some technical tools.

3.2.2 Lemma. *Let $A \in \mathrm{Sect}(\theta)$, $\theta \in [0, \pi)$ and $s \in (0, 1)$, then for all $\theta' > \theta$ there is a constant $c > 0$ such that*

$$\left\| \lambda^s A^{1-s} (\lambda + A)^{-1} \right\| \leq c$$

for all $\lambda \in \Sigma_{\pi-\theta'}$; if $A \in \mathrm{RSect}(\theta)$, then we have

$$\mathcal{R}\{\lambda^s A^{1-s}(\lambda + A)^{-1} : \lambda \in \Sigma_{\pi-\theta'}\} < \infty.$$

Proof. The first statement is an immediate consequence of the moment inequality [61, Prop. 2.12] or can alternatively be deduced from [66, Lemma 1.4.1]. The assertion on the R-boundedness follows readily from Lemma 1.10.6 or [80, Lemma 10]. □

We denote by V the space $\mathcal{D}(A^{1/2})$ equipped with the norm $\|x\|_V = \|(1 + A^{1/2})x\|$. Note that this norm is equivalent to the graph norm of $A^{1/2}$ as -1 is in the resolvent set of $A^{1/2}$. Observe that the two Banach spaces V and X are isomorphic (the isomorphism is just $(1 + A^{1/2})^{-1}$); hence, if X is a Hilbert space or a UMD Banach space, then so is V.

We consider the operator $\mathcal{A}$ on $V \times X$ given on its domain $\mathcal{D}(\mathcal{A}) = \mathcal{D}(A) \times V$ by the operator matrix

$$\mathcal{A} = \begin{pmatrix} 0 & I \\ A & 0 \end{pmatrix}. \tag{3.8}$$

The spectra of the operators A and $\mathcal{A}$ are closely connected: we have $\sigma(A) = \sigma(\mathcal{A})^2$.

3.2.3 Lemma. *Let A be sectorial. Then, for all $\lambda \in \mathbb{C}$ we have $\lambda \in \rho(\mathcal{A})$ if and only if $\lambda^2 \in \rho(A)$, i.e. $\sigma(A) = \sigma(\mathcal{A})^2$; in this case we have the relation*

$$R(\lambda, \mathcal{A}) = \begin{pmatrix} \lambda R(\lambda^2, A) & R(\lambda^2, A) \\ AR(\lambda^2, A) & \lambda R(\lambda^2, A) \end{pmatrix}.$$

In particular, $\lambda \in \rho(\mathcal{A}) \Leftrightarrow -\lambda \in \rho(\mathcal{A})$; moreover $\sigma(\mathcal{A}) \subset \overline{\Sigma_\theta}$ if and only if $\sigma(A) \subset \overline{\Sigma_{\alpha,\omega}}$ where $\theta \geq 0$, $\alpha = (0, \pi)$ and $\omega = (\theta/2, \theta/2)$.

Proof. If $\lambda^2 \in \rho(A)$, it is straightforward to verify that the operator on the right hand side is indeed the inverse of $\lambda - \mathcal{A}$. On the other hand, if $0 \neq \lambda \in \rho(\mathcal{A})$, let $P : V \times X \to X$ denote the coordinate projection and $\iota : X \to V \times X$ the canonical injection; then it is easy to check that $\lambda^{-1}PR(\lambda, \mathcal{A})\iota$ equals $R(\lambda^2, A)$ (at first for $x \in \mathcal{D}(A)$ and then by a density argument). If $0 \in \rho(\mathcal{A})$, then the resolvent $R(\lambda, \mathcal{A})$ is bounded in norm on a small neighborhood U about the origin. By what we have shown up to now, we deduce that the upper left entry $\lambda R(\lambda^2, A) \in \mathcal{L}(X)$ is bounded on $U \setminus \{0\}$. However, as $\|R(z, A)\| \geq (\mathrm{dist}(z, \sigma(A)))^{-1}$, this forces that $0 \in \rho(A)$. Indeed, if 0 was in the spectrum of A, we had the inequality $|\lambda|^{-1} \leq \|R(\lambda, \mathcal{A})\| \leq c|\lambda|^{-1/2}$ for all λ sufficiently small, a contradiction. □

Using this representation of the resolvent we may deduce the following statement on (R-)sectoriality.

3.2.4 Proposition. *Let A be sectorial and let $\theta > 0$, $\alpha = (0, \pi)$ and $\omega = (\theta/2, \theta/2)$. Then, for all $r > 0$, $\Lambda_A \subset \rho(A)$ if and only if $\Lambda_{\mathcal{A}} \subset \rho(\mathcal{A})$ where*

$$\Lambda_A = \{\lambda \in \mathbb{C} : |\lambda| \geq r^2\} \setminus \overline{\Sigma_\theta} \quad \text{and} \quad \Lambda_{\mathcal{A}} = \{\lambda \in \mathbb{C} : |\lambda| \geq r\} \setminus \overline{\Sigma_{\alpha,\omega}}.$$

Moreover, the set

$$\{\lambda R(\lambda, A) : \lambda \in \Lambda_A\}$$

is bounded (R-bounded) if and only if the set

$$\{\lambda R(\lambda, \mathcal{A}) : \lambda \in \Lambda_{\mathcal{A}}\}$$

is bounded (R-bounded).
   The operator $\mathcal{A}$ is (R-)i-bisectorial if and only if in addition $0 \in \rho(A)$.

Proof. First observe that Lemma 3.2.3 gives that the correspondence of the spectra is as claimed. Hence, it remains to establish the equivalence of the (R-)boundedness of the corresponding sets of operators.

   Making use of appropriate projections and embeddings one finds, as in the foregoing proof, that the resolvent operator $R(\lambda, \mathcal{A})$ is bounded if and only if the four entries are bounded. The uniform resolvent bounds required for the (R-)sectoriality of $\mathcal{A}$ are again equivalent to uniform bounds on each of the four entries. We will estimate each of them separately. Identifying V and X by the canonical isomorphism $(1 + A^{1/2})^{-1}$ we have to consider the set of operators

$$\lambda^2 R(\lambda^2, A), \quad (1 + A^{1/2})\lambda R(\lambda^2, A), \quad \lambda A R(\lambda^2, A)(1 + A^{1/2})^{-1}$$

for $\lambda \in \Lambda_{\mathcal{A}}$ and to show that they are (R-)bounded in $\mathcal{L}(X)$. This is obvious for the first operator; moreover, this yields also the converse of the statement. We write the second operator as a sum $\lambda R(\lambda^2, A) + A^{1/2}\lambda R(\lambda^2, A)$; the (R-)boundedness of the first term follows, as $|\lambda| \geq r$, from the (R-)sectoriality of A and the second one is bounded by Lemma 3.2.2. We may rewrite the third operator as $\lambda A^{1/2}R(\lambda^2, A)A^{1/2}(1 + A^{1/2})^{-1}$. The boundedness of the operator $A^{1/2}(1 + A^{1/2})^{-1}$ and Lemma 3.2.2 give together the required (R-)boundedness.

   It is clear, that the invertibility of A implies the invertibility of $\mathcal{A}$; hence there is no problem at zero since the required estimates at infinity were established above. Conversely, let $0 \in \sigma(A)$. Assume that $\mathcal{A}$ is bisectorial; i.e. the operator $\lambda R(\lambda, \mathcal{A})$ is norm bounded in $\mathcal{L}(V \times X)$ on $i\mathbb{R} \setminus \{0\}$. In particular, considering the upper right entry, we find that $c \geq \left\| (1 + A^{1/2})\lambda R(\lambda^2, A) \right\| \geq \|\lambda R(\lambda^2, A)\| - \left\| A^{1/2}\lambda R(\lambda^2, A) \right\|$ for all $\lambda \in i\mathbb{R} \setminus \{0\}$. Letting tend λ to zero, we obtain by Lemma 3.2.2 that the term we subtract remains bounded, whereas the other term blows up, as $\|\lambda R(\lambda^2, A)\|$ grows at least like $|\lambda|^{-1}$. This contradiction proves that A is invertible. $\square$

   The relation of the spectra is illustrated by the Figure 3.1.
   We will use the matrix $\mathcal{A}$ in order to prove L_p-maximal regularity on the line by reducing the second order equation to a first order system.

3.2.5 Theorem. *Let $A \in \mathrm{RSect}_d(\theta)$, $\theta \in [0, \pi)$, on the UMD Banach space X and assume that A is invertible. Let $1 < p < \infty$. Then for all $f \in L_p(\mathbb{R}, X)$ there is a unique $u \in W^{2,p}(\mathbb{R}, X) \cap W^{1,p}(\mathbb{R}, V) \cap L_p(\mathbb{R}, \mathcal{D}(A))$ such that $u'' - Au = f$ on $\mathbb{R}$.*

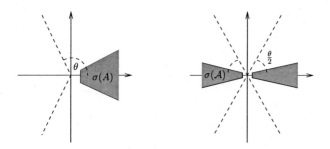

Figure 3.1: The relation between the spectral angles of A and $\mathcal{A}$. The position of the spectrum is indicated by the area hatched in grey.

Proof. We will first prove existence of a solution. Define $F = (0, f)^T \in L_p(\mathbb{R}, V \times X)$ and denote by $\mathcal{A}$ the operator matrix associated with A by (3.8). Then $\mathcal{A}$ is a densely defined R-bisectorial operator on the UMD space $V \times X$. More precisely, the set $\{isR(is, \mathcal{A}) : s \in \mathbb{R}\}$ is R-bounded. Hence we have, by the previously considered first order case, L_p-maximal regularity on the line for the first order problem

$$U' - \mathcal{A}U = F; \qquad (3.9)$$

that is there is a unique $U \in W^{1,p}(\mathbb{R}, V \times X) \cap L_p(\mathbb{R}, \mathcal{D}(\mathcal{A}))$ solving (3.9). Denoting the components of U by u_k, $k = 1, 2$, we find that $u_1 \in W^{1,p}(\mathbb{R}, V) \cap L_p(\mathbb{R}, \mathcal{D}(A))$ and that $u_1' = u_2 \in W^{1,p}(\mathbb{R}, X)$ whence $u_1 \in W^{2,p}(\mathbb{R}, X)$. Furthermore, $u_1'' = u_2' = Au_1 + f$ and $Au_1 = u_1'' - f \in L_p(\mathbb{R}, X)$. Hence u_1 is the desired solution.

The solution is unique; indeed, if $u \in W^{2,p}(\mathbb{R}, X) \cap W^{1,p}(\mathbb{R}, V) \cap L_p(\mathbb{R}, \mathcal{D}(A))$ is a solution, then put $U = (u, u')^T$ and $F = (0, f)^T$. Clearly, $U \in W^{1,p}(\mathbb{R}, V \times X) \cap L_p(\mathbb{R}, \mathcal{D}(\mathcal{A}))$ and $F \in L_p(\mathbb{R}, V \times X)$; it is straightforward to verify that U solves (3.9). Therefore, U is unique and consequently also u. □

For example, choosing $A = -\Delta + \epsilon$, $\epsilon > 0$, the negative translated Laplacian we obtain maximal regularity for the elliptic problem $u'' + \Delta u - \epsilon u = f$. The shift was necessary in order to have an invertible operator A. Another possible example is $A = -\Delta = -\Delta_\Omega^D$ the negative Laplacian on a bounded regular domain with Dirichlet boundary conditions.

The isomorphism $F \mapsto U$ yields the isomorphism $f \mapsto u$ between the Banach spaces $L_p(\mathbb{R}, X)$ and $W^{2,p}(\mathbb{R}, X) \cap W^{1,p}(\mathbb{R}, \mathcal{D}(A^{1/2})) \cap L_p(\mathbb{R}, \mathcal{D}(A))$. Combining this observation with Theorem 3.2.1 we find the following interpolatory identity.

3.2.6 Corollary. *Assume that X be a UMD Banach space and that A is a densely defined R-sectorial operator. Then*

$$W^{2,p}(\mathbb{R}, X) \cap L_p(\mathbb{R}, \mathcal{D}(A)) = W^{2,p}(\mathbb{R}, X) \cap W^{1,p}(\mathbb{R}, \mathcal{D}(A^{1/2})) \cap L_p(\mathbb{R}, \mathcal{D}(A)). \qquad (3.10)$$

We may assume A to be invertible, because $\mathcal{D}((A + \epsilon)^{1/2}) = \mathcal{D}(A^{1/2})$ for $\epsilon > 0$, see [61, Prop. 2.9].

3.2.7 Remarks. Observe that we only needed for the proof the fact the operator matrix $\mathcal{A}$ satisfies L_p-maximal regularity on the line for the first order problem; note also that we did not use the full strength of that result, as we only considered inhomogeneities of the form $F = (0, f)^T$.

If X is a UMD space and $B \in \text{Sect}_d(\theta)$, $\theta \in [0, \pi/2)$, then B satisfies maximal regularity for the first order problem if and only if A given by $-A = B^2 \in \text{Sect}_d(2\theta)$ satisfies maximal regularity for the second order problem [29]. This allows to characterize Hilbert spaces (among UMD spaces) by maximal regularity for the second order problem (compare Cor. 3.1.8).

In the following we will consider the operator matrix $\mathcal{A}$ in some more detail. In the same spirit as in the argument giving bisectoriality we find that if A has a bounded H^∞-functional calculus, then so does $\mathcal{A}$ with an analogous result concerning the angles. These results seem to be new. As we require $\mathcal{A}$ to be bisectorial, we have to assume that A (and thus also $\mathcal{A}$) are invertible. In this case we can choose the homogeneous norm $\|x\|_V = \|A^{1/2}x\|_X$ on V. In the following we will also assume A to be densely defined.

3.2.8 Theorem. *Let $A \in \text{Sect}(\theta)$, $\theta \in [0, \pi)$, be invertible and densely defined. For each $f \in H_0^\infty(\Sigma_{\alpha, \omega})$, $\alpha = (0, \pi)$ and $\omega' = (\theta'/2, \theta'/2)$, $\theta < \theta' < \pi$, the operator $f(\mathcal{A})$ is in $\mathcal{L}(V \times X)$ and given by the formula*

$$f(\mathcal{A}) = \frac{1}{2} \begin{pmatrix} g(A) + h(A) & A^{-1/2}(-g(A) + h(A)) \\ (-g(A) + h(A))A^{1/2} & g(A) + h(A) \end{pmatrix}, \qquad (3.11)$$

where $g, h \in H_0^\infty(\Sigma_{\theta'})$ are given by $g(z) = f(-\sqrt{z})$ and $h(z) = f(\sqrt{z})$.

Proof. Let $f \in H_0^\infty(\Sigma_{\alpha, \omega'})$ then

$$f(\mathcal{A}) = \frac{1}{2\pi i} \int_\Gamma f(\lambda) R(\lambda, \mathcal{A}) \, d\lambda$$

where the contour Γ is the boundary of $\Sigma_{\alpha, \tau}$ with $\tau = (\sigma/2, \sigma/2)$, $\theta < \sigma < \theta'$, equipped with the proper orientation. This integral converges in $\mathcal{L}(V \times X)$ as $\mathcal{A}$ is bisectorial. Denote the part of Γ lying in the left half-plane by Γ_l and the part in the right half-plane by Γ_r. Then

$$f(\mathcal{A}) = \frac{1}{2\pi i} \int_{\Gamma_l} f(\lambda) \begin{pmatrix} \lambda & I \\ A & \lambda \end{pmatrix} R(\lambda^2, A) \, d\lambda + \frac{1}{2\pi i} \int_{\Gamma_r} f(\lambda) \begin{pmatrix} \lambda & I \\ A & \lambda \end{pmatrix} R(\lambda^2, A) \, d\lambda$$

where $\begin{pmatrix} \lambda & I \\ A & \lambda \end{pmatrix} R(\lambda^2, A)$ means that we multiply each matrix entry with the resolvent $R(\lambda^2, A)$, i.e. all the occurring entries are actually still bounded operators. We calculate the two integrals by substitution. In the first integral we substitute $\lambda = -\sqrt{\mu}$; this is a conformal map of the left half-plane onto the cut plane $\mathbb{C} \setminus (-\infty, 0]$ under which the contour Γ_l corresponds to the admissible contour $\Gamma_\sigma = \partial\Sigma_\sigma$ for the sectorial operator A. In the second integral we make the substitution $\lambda = \sqrt{\mu}$. An easy calculation gives

$$f(\mathcal{A}) = \frac{1}{2\pi i} \int_{\Gamma_\sigma} \frac{1}{2} \begin{pmatrix} (f(-\sqrt{\mu}) + f(\sqrt{\mu})) & (f(\sqrt{\mu}) - f(-\sqrt{\mu}))\mu^{-1/2} \\ (f(\sqrt{\mu}) - f(-\sqrt{\mu}))\mu^{-1/2}A & (f(-\sqrt{\mu}) + f(\sqrt{\mu})) \end{pmatrix} R(\mu, A)\, d\mu.$$

It is easy to see that the upper left entry of the thus defined operator $f(\mathcal{A})$ is just $\frac{1}{2}(g(A) + h(A)) \in \mathcal{L}(X)$, where $g, h \in H_0^\infty(\Sigma_{\theta'})$ are given by $g(z) = f(-\sqrt{z})$ and $h(z) = f(\sqrt{z})$. Observe that $\|g\|_{H^\infty(\Sigma_{\theta'})} = \|h\|_{H^\infty(\Sigma_{\theta'})} = \|f\|_{H^\infty(\Sigma_{\alpha,\omega'})}$. It follows directly from Theorem 1.7.3 that the lower left entry equals $\frac{1}{2}(-g(A) + h(A))A^{1/2} \in \mathcal{L}(V, X)$. One proceeds analogously with the remaining two entries. Summarizing we find that

$$f(\mathcal{A}) = \frac{1}{2} \begin{pmatrix} g(A) + h(A) & A^{-1/2}(-g(A) + h(A)) \\ (-g(A) + h(A))A^{1/2} & g(A) + h(A) \end{pmatrix} \in \mathcal{L}(V \times X).$$

$\square$

We observe that the terms involving g stem from the integral over the left contour Γ_l, whereas the terms involving h arise from the contour Γ_r.

3.2.9 Proposition. *Let $A \in \text{Sect}(\theta)$, $\theta \in [0, \pi)$, be invertible, densely defined and assume that A has a bounded $H^\infty(\Sigma_{\theta'})$-functional calculus. Then $\mathcal{A}$ has a bounded $H^\infty(\Sigma_{\alpha,\omega'})$-functional calculus, where $\alpha = (0, \pi)$ and $\omega' = (\theta'/2, \theta'/2)$. Moreover, for $f \in H^\infty(\Sigma_{\alpha,\omega})$ the bounded operator $f(\mathcal{A})$ is given by the identity (3.11).*

Proof. Let $f \in H_0^\infty(\Sigma_{\alpha,\omega})$. Estimating the four entries separately we find that the norm of each entry is bounded by $C \|f\|_{H^\infty(\Sigma_{\alpha,\omega'})}$, where C is the norm of bounded $H^\infty(\Sigma_{\theta'})$-functional calculus for A; for example the norm of the lower left entry is estimated as follows:

$$\begin{aligned}
\frac{1}{2} \left\| A^{1/2}(-g(A) + h(A)) \right\|_{\mathcal{L}(V,X)} &= \frac{1}{2} \left\| -g(A) + h(A) \right\|_{\mathcal{L}(X)} \\
&\leq \frac{1}{2}(\|g(A)\| + \|h(A)\|) \\
&\leq \frac{1}{2}C \|g\|_{H^\infty(\Sigma_{\theta'})} + \frac{1}{2}C \|h\|_{H^\infty(\Sigma_{\theta'})} = C \|f\|_{H^\infty(\Sigma_{\alpha,\omega'})}.
\end{aligned}$$

Hence, the $H^\infty(\Sigma_{\alpha,\omega'})$-functional calculus for $\mathcal{A}$ is also bounded with norm at most $2C$.

Note that $\mathcal{A}$ is densely defined since we assume A to be densely defined. Let ψ_n be a sequence of regularizers for $\mathcal{A}$. Considering the sequence $f_n = f\psi_n \in H_0^\infty(\Sigma_{\alpha,\omega})$ we obtain the validity of formula (3.11) for general f. $\square$

More is true; it is a consequence of the Convergence Lemma and Remark 1.4.19 that the operator $f(\mathcal{A})$ is bounded if the operators $g(A)$ and $h(A)$ are bounded operators in $\mathcal{L}(X)$. Choosing for $f = p = 1_{\{\mathcal{R}e(z) > 0\}}$ we find that $g = 0$ and $h \equiv 1$, hence $g(A) = 0$ and $h(A) = I$. We summarize these considerations.

3.2.10 Proposition. *Let $A \in \text{Sect}(\theta)$, $\theta \in [0,\pi)$, be invertible and densely defined. The spectral projection $P = p(\mathcal{A})$ corresponding to the right half-plane is bounded in $\mathcal{L}(V \times X)$ and given by*

$$P = \frac{1}{2}\begin{pmatrix} I & A^{-1/2} \\ A^{1/2} & I \end{pmatrix}.$$

The spectral decomposition is given by $X_1 = \mathcal{R}(P) = \ker I - P = \{(x,y) \in V \times X : y = A^{1/2}x\}$ and $X_2 = \ker P = \{(x,y) \in V \times X : y = -A^{1/2}x\}$.

Let $\mathcal{A}_i$ denote the restriction of $\mathcal{A}$ to X_i. The resolvent $R(\lambda, \mathcal{A}_1)$ is obtained by inserting $(\lambda - \cdot)^{-1}p$ into the functional calculus for $\mathcal{A}$. For $\lambda \in \rho(\mathcal{A}_1) = \rho(A^{1/2})$, the formula (3.11) proven above gives that

$$R(\lambda, \mathcal{A}_1) = \frac{1}{2}\begin{pmatrix} R(\lambda, A^{1/2}) & A^{-1/2}R(\lambda, A^{1/2}) \\ R(\lambda, A^{1/2})A^{1/2} & R(\lambda, A^{1/2}) \end{pmatrix}.$$

The resolvent of $\mathcal{A}_2$ may be obtained by considering the function $(\lambda - \cdot)^{-1}(1 - p)$. We obtain for $\lambda \in \rho(\mathcal{A}_2) = -\rho(A^{1/2})$

$$R(\lambda, \mathcal{A}_2) = \frac{1}{2}\begin{pmatrix} -R(-\lambda, A^{1/2}) & A^{-1/2}R(-\lambda, A^{1/2}) \\ R(-\lambda, A^{1/2})A^{1/2} & -R(-\lambda, A^{1/2}) \end{pmatrix}.$$

Of course, using these formulae for the resolvents it is easy to verify directly that $X = X_1 \oplus X_2$ is indeed the desired spectral decomposition.

3.2.11 Example. Let X be a Banach space. The operator $A = \alpha I$, $\alpha > 0$, is sectorial and invertible with spectrum $\sigma(A) = \{\alpha\}$. The square-root $A^{1/2}$ is the operator $\sqrt{\alpha}I$ with domain $V = X$. Hence, the operator $\mathcal{A} = \begin{pmatrix} 0 & I \\ \alpha I & 0 \end{pmatrix}$ defined on $X \times X$ is bisectorial with spectrum $\sigma(\mathcal{A}) = \{-\sqrt{\alpha}, \sqrt{\alpha}\}$. The bounded spectral projection is given by the operator matrix

$$P = \frac{1}{2}\begin{pmatrix} I & \frac{1}{\sqrt{\alpha}} \\ \sqrt{\alpha} & I \end{pmatrix};$$

the corresponding spectral decomposition is given by

$$X \times X = \{(x, \sqrt{\alpha}x) : x \in X\} \oplus \{(x, -\sqrt{\alpha}x) : x \in X\}.$$

Considering the lower right entry in formula (3.11) yields that if $\mathcal{A}$ has a bounded functional calculus then so does A (we may choose g to be zero by multiplying with p). This gives the following characterization.

3.2.12 Corollary. *Let $A \in \text{Sect}(\theta)$, $\theta \in [0,\pi)$, be invertible and densely defined. Then A has a bounded $H^\infty(\Sigma_{\theta'})$-functional calculus if and only if $\mathcal{A}$ has a bounded $H^\infty(\Sigma_{\alpha,\omega'})$-functional calculus, where $\alpha = (0,\pi)$ and $\omega' = (\theta'/2, \theta'/2)$.*

3.3   Quasilinear equations

We will apply now maximal regularity on the line to obtain solutions for quasilinear problems on the line. First we will formulate the results in an abstract setting. Then these will be applied to various examples of concrete partial differential equations. The results of this section seem to be new.

3.3.1   The abstract theorems

The smoothing property of maximal regularity will allow to apply Banach's fixed point theorem provided we have a strict contraction. In [27], [30] and [103] were considered quasilinear equations of first and second order on a finite interval. There the existence of local solutions was proved by making the length of the interval small enough. In our case, we are looking for global solutions and we achieve the contraction property by imposing the requirement that the norm of the inhomogeneity be sufficiently small.

First order equation.   In the following we will study the existence and uniqueness of global solutions to the first order problem

$$u'(t) + A(u(t))u(t) + F(t, u(t)) = f(t), \qquad t \in \mathbb{R} \tag{3.12}$$

on a Banach space X.

Let D_A be a Banach space that is densely and continuously embedded into X; let $p \in (1, \infty)$. Define the *maximal regularity space*

$$\mathrm{MR}_p(X, D_A) = W^{1,p}(\mathbb{R}, X) \cap L^p(\mathbb{R}, D_A)$$

equipped with the norm given by

$$\|u\|_{\mathrm{MR}_p} = \|u\|_{W^{1,p}(\mathbb{R}, X)} + \|u\|_{L^p(\mathbb{R}, D_A)}$$

and the *trace space*

$$\mathrm{Tr}_p(X, D_A) = \{u(0) : u \in \mathrm{MR}_p(X, D_A)\}$$

which becomes a Banach space endowed with the norm

$$\|u_0\|_{\mathrm{Tr}_p} = \inf\{\|u\|_{\mathrm{MR}_p} : u \in \mathrm{MR}_p \text{ and } u(0) = u_0\}.$$

In fact, the characterization of real interpolation spaces as trace spaces (see [89, p. 20]) yields at once

3.3.1 Lemma. *Let $p \in (1, \infty)$ and let $p^* = \frac{p}{p-1}$ denote the conjugate exponent of p. Then*

$$\mathrm{Tr}_p = (X, D_A)_{\frac{1}{p^*}, p}.$$

The following lemma shows, that the space MR_p embeds continuously into $C_0(\mathbb{R}, \mathrm{Tr}_p)$.

3.3.2 Lemma. *Let $u \in \mathrm{MR}_p$, then $u \in C_0(\mathbb{R}, \mathrm{Tr}_p)$ and*

$$\|u(t)\|_{\mathrm{Tr}_p} \leq \|u\|_{\mathrm{MR}_p}, \qquad t \in \mathbb{R}. \tag{3.13}$$

Proof. Let $u \in \mathrm{MR}_p$, for $t \in \mathbb{R}$ define the translates $v_t(s) = u(s + t)$, then $v_t \in \mathrm{MR}_p$ with $\|v_t\|_{\mathrm{MR}_p} = \|u\|_{\mathrm{MR}_p}$. Hence, $u(t) = v_t(0) \in \mathrm{Tr}_p$ and $\|u(t)\|_{\mathrm{Tr}_p} \leq \|v_t\|_{\mathrm{MR}_p} = \|u\|_{\mathrm{MR}_p}$. The continuity of the map $t \mapsto u(t)$ is an immediate consequence of the continuity of the translation group on L_p. Indeed, this implies the continuity of the map $t \mapsto v_t$ from $\mathbb{R}$ into MR_p, hence $\|u(t) - u(t_0)\|_{\mathrm{Tr}_p} \leq \|v_t - v_{t_0}\|_{\mathrm{MR}_p} \to 0$ for $t \to t_0$.

It remains to prove that u vanishes at infinity. To this end let $\phi : \mathbb{R} \to \mathbb{R}$ be a test-function satisfying $\phi(t) = 1$ for $|t| \leq 1$ and $\mathrm{supp}\,\phi \subset (-2, 2)$. For $n \in \mathbb{Z}$ denote by ϕ_n the translated function $\phi_n(t) = \phi(t - n)$. The functions $\phi_n u$ are elements of MR_p and there is a constant C_ϕ (uniform in $n \in \mathbb{Z}$) such that $\|\phi_n u\|_{\mathrm{MR}_p} \leq C_\phi \|u\|_{\mathrm{MR}_p(\mathrm{supp}\,\phi_n)}$, where $\|u\|_{\mathrm{MR}_p(\mathrm{supp}\,\phi_n)}$ is given by $\|u\|_{W^{1,p}(\Omega_n, X)} + \|f\|_{L_p(\Omega_n, X)}$ with $\Omega_n = \{t \in \mathbb{R} : \phi_n(t) > 0\}$. For $|n| \to \infty$ the right hand side converges to zero, since $u \in \mathrm{MR}_p$. Given $t \in (n - 1, n + 1)$ we have $\|u(t)\|_{\mathrm{Tr}_p} = \|(\phi_n u)(t)\|_{\mathrm{Tr}_p} \leq \|\phi_n u\|_{\mathrm{MR}_p}$, which proves that $u \in C_0(\mathbb{R}, \mathrm{Tr}_p)$. $\qquad\square$

Given $r \in [0, \infty)$, let U_r denote the closed r-ball about zero in the trace space Tr_p, that is $U_r = \{u \in \mathrm{Tr}_p : \|u\|_{\mathrm{Tr}_p} \leq r\}$. We will consider a family of operators $A : U_r \to \mathcal{L}(D_A, X)$. For simplicity, let $A_0 = A(0)$; observe that the constant null function is an element of MR_p, hence $0 \in \mathrm{Tr}_p$.

3.3.3 Theorem. *Let $p \in (1, \infty)$ and assume that the linear problem*

$$u' + A_0 u = g, \tag{3.14}$$

has L_p-maximal regularity, that is for all $g \in L_p(\mathbb{R}, X)$ there is a unique element $u \in \mathrm{MR}_p$ solving (3.14). Let M denote the norm of the corresponding solution operator $g \mapsto u$.

Assume moreover, that there is $r_0 \in (0, \infty)$ such that

1. *the function $A : U_{r_0} \to \mathcal{L}(D_A, X)$ is Lipschitz continuous with constant L, i.e.*

$$\|A(v) - A(w)\|_{\mathcal{L}(D_A, X)} \leq L \|v - w\|_{\mathrm{Tr}_p};$$

2. *the function $F : \mathbb{R} \times U_{r_0} \to X$ is continuous and Lipschitz continuous with respect to the second variable, more precisely we require the estimates*

$$\|F(t, v)\|_X \leq h_1(t) \|v\|_{\mathrm{Tr}_p} \tag{3.15}$$

$$\|F(t, v) - F(t, w)\|_X \leq h_2(t) \|v - w\|_{\mathrm{Tr}_p} \tag{3.16}$$

for all $v, w \in U_{r_0}$, $t \in \mathbb{R}$, where $h_1, h_2 \in L_p(\mathbb{R})$.

Assume that $\|h_1\|_{L_p} < M^{-1}$, $\|h_2\|_{L_p} < M^{-1}$; then choose $0 \leq r \leq r_0$ such that $LMr + M \|h_1\|_{L_p} < 1$ and $2LMr + M \|h_2\|_{L_p} < 1$. Let $f \in L_p(\mathbb{R}, X)$ such that $\|f\|_{L_p(\mathbb{R}, X)} \leq r(M^{-1} - Lr - \|h_1\|_{L_p})$. Then the quasilinear problem (3.12) admits a unique solution $u \in \{v \in \mathrm{MR}_p : \|v\|_{\mathrm{MR}_p} \leq r\}$.

Proof. The argument will be based on Banach's fixed point theorem. As we have maximal regularity for the problem (3.14), the boundedness of the associated solution operator implies the inequality

$$\|u\|_{\mathrm{MR}_p} \le M \|g\|_{L_p(\mathbb{R},X)}$$

where u is the unique solution of the linear Cauchy problem $u' + A_0 u = g$ on $\mathbb{R}$.

Denote by $\mathcal{C}$ the closed r-ball about zero in MR_p, that is $\mathcal{C} = \{u \in \mathrm{MR}_p : \|u\|_{\mathrm{MR}_p} \le r\}$. We denote by Φ the map that associates to a given $v \in \mathcal{C}$ the unique solution u of (3.14) for the inhomogeneity $g = (A_0 - A(v))v - F(\cdot, v) + f$. We claim that $g \in L_p(\mathbb{R}, X)$ and that $\Phi : \mathcal{C} \to \mathcal{C}$. Indeed, we have the following estimates:

$$\|\Phi(v)\|_{\mathrm{MR}_p} \le M \|(A_0 - A(v))v - F(\cdot, v) + f\|_{L_p(\mathbb{R},X)}$$
$$\le M(\|A_0 - A(v)\|_{C_b(\mathbb{R}, \mathcal{L}(D_A, X))} \|v\|_{L_p(\mathbb{R}, D_A)} + \|F(\cdot, v)\|_{L_p(\mathbb{R},X)} + \|f\|_{L_p(\mathbb{R},X)}).$$

The map $A_0 - A(v) : \mathbb{R} \to \mathcal{L}(D_A, X)$ is continuous, because v is in $C_0(\mathbb{R}, \mathrm{Tr}_p)$ with range in U_r by the previous lemma and because A is even Lipschitz continuous on U_{r_0}. The boundedness is again a consequence of the Lipschitz continuity: using the Lipschitz continuity and applying then pointwise Lemma 3.3.2 on the first term and estimating the L_p-norm involving F by means of (3.15) we find the upper estimate

$$\le M(L \|v\|_{\mathrm{MR}_p} \|v\|_{L_p(\mathbb{R}, D_A)} + \|h_1\|_{L_p} \|v\|_{\mathrm{MR}_p} + \|f\|_{L_p(\mathbb{R},X)})$$
$$\le MLr^2 + M \|h_1\|_{L_p} r + M \|f\|_{L_p(\mathbb{R},X)} \le r,$$

which proves the claim.

Next, we show that the mapping Φ is a strict contraction on the set $\mathcal{C}$. Let $v, \bar{v} \in \mathcal{C}$, then, as $\Phi(v) - \Phi(\bar{v})$ is the unique solution of (3.14) corresponding to the inhomogeneity $(A_0 - A(v))v - (A_0 - A(\bar{v}))\bar{v} + F(\cdot, \bar{v}) - F(\cdot, v)$, we find, estimating as above,

$$\|\Phi(v) - \Phi(\bar{v})\|_{\mathrm{MR}_p}$$
$$\le M \|(A_0 - A(\bar{v}))(v - \bar{v}) + (A(\bar{v}) - A(v))v + (F(\cdot, \bar{v}) - F(\cdot, v))\|_{L_p(\mathbb{R},X)}$$
$$\le ML(\|\bar{v}\|_{\mathrm{MR}_p} \|v - \bar{v}\|_{L_p(\mathbb{R}, D_A)} + \|v - \bar{v}\|_{\mathrm{MR}_p} \|v\|_{L_p(\mathbb{R}, D_A)}) + M \|h_2\|_{L_p} \|v - \bar{v}\|_{\mathrm{MR}_p}$$
$$\le (2LMr + M \|h_2\|_{L_p}) \|v - \bar{v}\|_{\mathrm{MR}_p}.$$

By Banach's fixed point theorem the strict contraction Φ has a unique fixed point u in $\mathcal{C}$. This u is the desired solution of equation (3.12). $\qquad\square$

3.3.4 Remarks.   1. One may replace condition (3.15) by the estimate $\|F(t, v)\|_X \le h_1(t)(1 + \|v\|_{\mathrm{Tr}_p})$ if one adapts the condition on h_1 and f accordingly.

2. For a given f one may find an $\epsilon > 0$ such that the function $f 1_{[-\epsilon, \epsilon]}$ satisfies the required estimate.

3. Note that we may replace $A(\cdot)$ by $A(\cdot) + \lambda$ for $\lambda \in \mathbb{C}$ if we can guarantee that $A_0 + \lambda$ still satisfies maximal regularity. This does not change the continuity properties of A. However, the constant M will depend on λ.

Second order equation. In this paragraph we will consider the second order quasi-linear problem

$$u'' + A(u, u')u + F(t, u, u') = f(t), \qquad t \in \mathbb{R} \tag{3.17}$$

on a Banach space X. A related problem was already considered by Chill and Srivastava in [27]; there the authors studied the local existence of solutions of a related initial value problem. The setting here is different, but the methods employed are similar. The terminology introduced here is consistent with the one used in [27].

Let D_A be a Banach space that is densely and continuously embedded into X; let $p \in (1, \infty)$. Define the *maximal regularity space*

$$\mathrm{MR}_p(X, D_A) = W^{2,p}(\mathbb{R}, X) \cap L^p(\mathbb{R}, D_A)$$

equipped with the norm given by

$$\|u\|_{\mathrm{MR}_p} = \|u\|_{W^{2,p}(\mathbb{R}, X)} + \|u\|_{L^p(\mathbb{R}, D_A)}$$

and the *trace space*

$$\mathrm{Tr}_p(X, D_A) = \{(u(0), u'(0)) : u \in \mathrm{MR}_p(X, D_A)\}$$

which becomes a Banach space endowed with the norm

$$\|(u_0, u_1)\|_{\mathrm{Tr}_p} = \inf\{\|u\|_{\mathrm{MR}_p} : u \in \mathrm{MR}_p \text{ and } u(0) = u_0, u'(0) = u_1\}.$$

Using the characterization of real interpolation spaces ([89]) by means of trace spaces it is easy to see that the following inclusion is valid.

3.3.5 Lemma. *Let $p \in (1, \infty)$ and let $p^* = \frac{p}{p-1}$ denote the conjugate exponent of p. Then*

$$\mathrm{Tr}_p \subset (X, D_A)_{\frac{1}{p^*}, p} \times X$$

with continuous embedding.

Proof. Let $u \in \mathrm{MR}_p$, then in particular

$$u \in W^{1,p}(\mathbb{R}, X) \cap L_p(\mathbb{R}, D_A) \text{ and}$$
$$u' \in W^{1,p}(\mathbb{R}, X);$$

the claim now follows from [89, p. 20]. □

The interpolatory identity (3.10) implies the following sharper result which will be used in later applications.

3.3.6 Lemma. *Let X be a UMD space and A be a densely defined R-sectorial operator. If the common domain is given by $D_A = \mathcal{D}(A)$, we have the continuous embedding*

$$\mathrm{Tr}_p \subset (X, \mathcal{D}(A))_{\frac{1}{2p^*} + \frac{1}{2}, p} \times (X, \mathcal{D}(A^{1/2}))_{\frac{1}{p^*}, p}. \tag{3.18}$$

Proof. Let $D' = \mathcal{D}(A^{1/2})$ and observe that we have $u \in W^{1,p}(\mathbb{R}, D') \cap L_p(\mathbb{R}, \mathcal{D}(A))$ and $u' \in W^{1,p}(\mathbb{R}, X) \cap L_p(\mathbb{R}, D')$ by (3.10). Applying the identity for real interpolation spaces between domains of fractional powers of sectorial operators [115, Section 1.5.14], for $\theta = 1/p^* \in (0, 1)$, we obtain

$$(\mathcal{D}(A^{1/2}), \mathcal{D}(A))_{\theta,p} = (X, \mathcal{D}(A))_{\frac{1+\theta}{2},p},$$

which completes the proof.   □

Again, the space MR_p embeds continuously into $C_0(\mathbb{R}, \mathrm{Tr}_p)$.

3.3.7 Lemma. *Let $u \in \mathrm{MR}_p$, then $(u, u') \in C_0(\mathbb{R}, \mathrm{Tr}_p)$ and*

$$\|(u(t), u'(t))\|_{\mathrm{Tr}_p} \leq \|u\|_{\mathrm{MR}_p}, \qquad t \in \mathbb{R}. \tag{3.19}$$

Proof. The proof of Lemma 3.3.2 carries over to this setting requiring only minor notational modifications.   □

Given $r \in [0, \infty)$, let U_r denote the closed r-ball about zero in Tr_p, that is $U_r = \{u \in \mathrm{Tr}_p : \|u\|_{\mathrm{Tr}_p} \leq r\}$. Again, we consider a family of operators $A : U_r \to \mathcal{L}(D_A, X)$. For simplicity, let $A_0 = A(0, 0)$; observe that the constant null function is an element of MR_p, hence $0 \in \mathrm{Tr}_p$.

3.3.8 Theorem. *Let $p \in (1, \infty)$ and assume that the linear problem*

$$u'' + A_0 u = g, \tag{3.20}$$

has L_p-maximal regularity, that is for all $g \in L_p(\mathbb{R}, X)$ there is a unique element $u \in \mathrm{MR}_p$ solving (3.20). Let M denote the norm of the corresponding solution operator $g \mapsto u$.
Assume moreover, that there is $r_0 \in (0, \infty)$ such that

1. *the function $A : U_{r_0} \to \mathcal{L}(D_A, X)$ is Lipschitz continuous with constant L, i.e.*

$$\|A(v_1, v_2) - A(w_1, w_2)\|_{\mathcal{L}(D_A, X)} \leq L \|(v_1, v_2) - (w_1, w_2)\|_{\mathrm{Tr}_p};$$

2. *the function $F : \mathbb{R} \times U_{r_0} \to X$ is continuous and Lipschitz continuous with respect to the second variable. More precisely we require the estimates*

$$\|F(t, v_1, v_2)\|_X \leq h_1(t) \|(v_1, v_2)\|_{\mathrm{Tr}_p}$$
$$\|F(t, v_1, v_2) - F(t, w_1, w_2)\|_X \leq h_2(t) \|(v_1, v_2) - (w_1, w_2)\|_{\mathrm{Tr}_p}$$

for all $v = (v_1, v_2), w = (w_1, w_2) \in U_{r_0}$, $t \in \mathbb{R}$, where $h_1, h_2 \in L_p(\mathbb{R})$.

Assume that $\|h_1\|_{L_p} < M^{-1}$, $\|h_2\|_{L_p} < M^{-1}$; choose $0 \leq r \leq r_0$ such that $LMr + M \|h_1\|_{L_p} < 1$ and $2LMr + M \|h_2\|_{L_p} < 1$; assume that $\|f\|_{L_p(\mathbb{R}, X)} \leq r(M^{-1} - Lr - \|h_1\|_{L_p})$ then the problem (3.17) admits a unique solution $u \in \{v \in \mathrm{MR}_p : \|v\|_{\mathrm{MR}_p} \leq r\}$.

Proof. It is easy to see that the proof of the first order case is also valid in this case; there are only a few obvious changes required.   □

3.3.2   Applications to quasilinear equations on $\mathbb{R}$

Multiplicative perturbations of the Laplacian.   We will consider elliptic quasilinear equations of the form $u_{tt} + A(u)u = g$, where $A(u)$ is a multiplicative perturbation of the Laplacian. First we will consider bounded smooth domains and then arbitrary bounded open sets.

Smooth domains.   Let Ω be a bounded smooth domain in $\mathbb{R}^N$, and consider the elliptic quasilinear equation on a cylindrical domain

$$\begin{cases} u_{tt} + a(\|\nabla u\|_{L_2})\Delta u = g, & (t,x) \in \mathbb{R} \times \Omega, \\ u(t,x) = 0, & (t,x) \in \mathbb{R} \times \partial\Omega, \end{cases} \tag{3.21}$$

Here, the function $a : \mathbb{R}_+ \to \mathbb{R}$ is continuous, $a(0) > 0$, and g belongs to $L_q(\mathbb{R}, L_p(\Omega))$ with $2 \le q \le p < \infty$. The gradient acts only on the space variables.

3.3.9 Theorem. *Assume that the function a is Lipschitz continuous on bounded intervals. Then, for $g \in L_q(\mathbb{R}, L_p(\Omega))$ sufficiently small, there is a solution u of the equation $u_{tt} + a(\|\nabla u\|_{L_2})\Delta u = g$ satisfying*

$$u \in W^{2,q}(\mathbb{R}, L_p(\Omega)) \cap L_q(\mathbb{R}, W^{2,p}(\Omega) \cap \overset{\circ}{W}^{1,p}(\Omega)).$$

Moreover, $u \in C_0(\mathbb{R}, B_{pq}^{2/q^*})$.

The solution is unique in a ball about the origin of sufficiently small radius.

Proof. We define the Laplacian with Dirichlet boundary conditions on the UMD space $X = L_p(\Omega)$ by

$$\mathcal{D}(B) = W^{2,p}(\Omega) \cap \overset{\circ}{W}^{1,p}(\Omega),$$
$$Bu = \Delta u.$$

It is well known that $-B$ is a densely defined invertible sectorial operator of angle 0 that admits a bounded H^∞-functional calculus [11, Thm. 5.7], its angle is 0. Since $L_p(\Omega)$ has property UMD, the operator B is R-sectorial of angle 0 [74, Thm. 5.3]. By Theorem 3.2.1 the problem

$$u'' + \alpha\Delta u = f, \qquad t \in \mathbb{R}$$

has L_q-maximal regularity for all $\alpha > 0$. The associated maximal regularity space is

$$\mathrm{MR}_q = W^{2,q}(\mathbb{R}, L_p(\Omega)) \cap L_q(\mathbb{R}, W^{2,p}(\Omega) \cap \overset{\circ}{W}^{1,p}(\Omega))$$

and, by Lemma 3.3.5 and [116, p. 204], the associated trace space is continuously embedded in

$$\mathrm{Tr}_q \subset (B_{pq}^{2/q^*}(\Omega) \cap \overset{\circ}{B}_{pq}^{1/q^*}(\Omega)) \times L_p(\Omega),$$

where $q^* = q/(q-1)$.

In the abstract setting the problem (3.21) corresponds to $D_A = \mathcal{D}(B)$ and the operator family $A : \mathrm{Tr}_q \to \mathcal{L}(D_A, X)$, $A(v, w) = A(v) = a(\|\nabla v\|_{L_2})\Delta$, which depends actually only on the first coordinate. It remains to verify that A is Lipschitz continuous on bounded sets.

By the choice of p and q and the embedding of Besov spaces [115, Section 4.6.1] we have the following chain of continuous embeddings:

$$B_{pq}^{2/q^*} \hookrightarrow B_{pp}^{2/q^*} = W^{2/q^*, p} \hookrightarrow W^{1,p} \hookrightarrow W^{1,2}. \tag{3.22}$$

Hence, we have in particular the continuous inclusion $\mathrm{Tr}_q \subset W^{1,2}(\Omega) \times L_p(\Omega)$. Using this fact, we now show that A defined above is Lipschitz continuous on each bounded subset U of Tr_q. Let $\tilde{U}$ denote the image of U under the coordinate projection $\mathrm{Tr}_q \to B_{pq}^{2/q^*}(\Omega) \cap \mathring{B}_{pq}^{1/q^*}(\Omega)$. Observe that by (3.22) the numbers $\|\nabla v\|_{L_2}$, $v \in \tilde{U}$, are uniformly bounded, hence a is Lipschitz continuous on the bounded set $\{\|\nabla v\|_{L_2} : v \in \tilde{U}\}$. Note that, since Δ is invertible, we have the equivalence of norms $\|\Delta \cdot\|_{L_p} \sim \|\cdot\|_{W^{2,p}}$. Let $(v_1, v_2), (w_1, w_2) \in U$ with norm bounded by $r_0 > 0$ and $u \in D_A$, then

$$\begin{aligned}
\|(A(v_1, v_2) - A(w_1, w_2))u\|_X &= \left\|(a(\|\nabla v_1\|_{L_2}) - a(\|\nabla w_1\|_{L_2}))\Delta u\right\|_{L_p} \\
&\leq L \left|\|\nabla v_1\|_{L_2} - \|\nabla w_1\|_{L_2}\right| \|\Delta u\|_{L_p} \\
&\leq L \|\nabla(v_1 - w_1)\|_{L_2} \|u\|_{D_A} \\
&\leq L \|v_1 - w_1\|_{W^{1,2}} \|u\|_{D_A} \\
&\leq CL \|(v_1, v_2) - (w_1, w_2)\|_{\mathrm{Tr}_q} \|u\|_{D_A}.
\end{aligned}$$

Since $A(0) = \alpha\Delta$ with $\alpha = a(0) > 0$, the linear operator $A(0)$ has maximal regularity and hence the claim follows from an application of Theorem 3.3.8. $\qquad\square$

3.3.10 Remark. We may improve the regularity of the solution by an application of Lemma 3.3.6. Making use of the well known interpolation results for Besov spaces [115] we obtain that the first factor of the trace space Tr_q embeds into $B_{pq}^{1+1/q^*}(\Omega) \cap \mathring{B}_{pq}^{1/2+1/(2q^*)}(\Omega)$.

We modify the perturbation. Let Ω be a bounded smooth domain in $\mathbb{R}^N$, and consider the equation

$$\begin{cases} u_{tt} + a(u)\Delta u = g, & (t, x) \in \mathbb{R} \times \Omega, \\ u(t, x) = 0, & (t, x) \in \mathbb{R} \times \partial\Omega. \end{cases} \tag{3.23}$$

Here, the continuous function $a : \mathbb{C} \to \mathbb{R}$, $a(0) > 0$, acts pointwise on u and g belongs to $L_q(\mathbb{R}, L_p(\Omega))$ with $2 \leq q \leq p < \infty$ and $p > N$.

3.3.11 Theorem. *Assume that the function a is Lipschitz continuous on bounded sets. Then, for $g \in L_q(\mathbb{R}, L_p(\Omega))$ sufficiently small, there is a solution of (3.23) satisfying*

$$u \in W^{2,q}(\mathbb{R}, L_p(\Omega)) \cap L_q(\mathbb{R}, W^{2,p}(\Omega) \cap \mathring{W}^{1,p}(\Omega)).$$

Moreover, $u \in C_0(\mathbb{R}, B_{pq}^{2/q^*})$.

Proof. The argument goes along the same lines as before; for this reason we will omit some details. The operator family is given this time by $A : \mathrm{Tr}_q \to \mathcal{L}(D_A, X)$, $A(v, w) = A(v) = a(v)\Delta$. Observe that $\mathcal{D}(m\Delta) = \mathcal{D}(\Delta) = D_A$ for all $m \in L_\infty(\Omega)$. It suffices to prove Lipschitz continuity; we use the same notation as before. By the Sobolev embedding theorem [21, IX.14] we have the continuous inclusion $W^{1,p}(\Omega) \hookrightarrow L_\infty(\Omega)$ since $p > N$; hence, if $U \subset \mathrm{Tr}_q$ is a bounded set, we find by (3.22) for $(v_1, v_2), (w_1, w_2) \in U$ the estimate

$$
\begin{aligned}
\|(A(v_1, v_2) - A(w_1, w_2))u\|_X &= \|(a(v_1) - a(w_1))\Delta u\|_{L_p} \\
&\leq \|a(v_1) - a(w_1)\|_{L_\infty} \|\Delta u\|_{L_p} \\
&\leq L \|v_1 - w_1\|_{L_\infty} \|u\|_{D_A} \\
&\leq CL \|v_1 - w_1\|_{W^{1,p}} \|u\|_{D_A} \\
&\leq C'L \|(v_1, v_2) - (w_1, w_2)\|_{\mathrm{Tr}_q} \|u\|_{D_A} \,,
\end{aligned}
$$

which proves the claim.   □

The operators $-A(v)$ that occurred in the examples above are densely defined, invertible and R-sectorial of angle 0. Hence, we have maximal regularity for the linearized first order equations, and the corresponding quasilinear parabolic problems $u' \pm a(u)\Delta = g$, $u(t, x) = 0$ on the boundary, admit a solution $u \in W^{1,p}(\mathbb{R}, X) \cap L_p(\mathbb{R}, D_A)$, if the smallness conditions are satisfied. We state this for reference.

3.3.12 Theorem. *Let $\Omega \subset \mathbb{R}^N$ be a bounded smooth domain. Let $a : \mathbb{C} \to \mathbb{R}$, $a(0) > 0$, be Lipschitz continuous on bounded intervals. If $g \in L_q(\mathbb{R}, L_p(\Omega))$ is small enough, where $2 \leq q \leq p < \infty$,*

1. *then there is a solution $u \in W^{1,q}(\mathbb{R}, L_p(\Omega)) \cap L_q(\mathbb{R}, W^{2,p}(\Omega) \cap \mathring{W}^{1,p}(\Omega)) \cap C_0(\mathbb{R}, B_{pq}^{2/q^*})$ of the equation*

$$
u_t \pm a(\|\nabla u\|_{L_2})\Delta u = g, \qquad (t, x) \in \mathbb{R} \times \Omega
$$

   with Dirichlet boundary condition.

2. *and if $p > N$, then there is a solution $u \in W^{1,q}(\mathbb{R}, L_p(\Omega)) \cap L_q(\mathbb{R}, W^{2,p}(\Omega) \cap \mathring{W}^{1,p}(\Omega)) \cap C_0(\mathbb{R}, B_{pq}^{2/q^*})$ of the equation*

$$
u_t \pm a(u)\Delta u = g, \qquad (t, x) \in \mathbb{R} \times \Omega
$$

   with Dirichlet boundary condition.

The boundary condition is incorporated in the domain of the Laplacian.

General domains. Let $\Omega \subset \mathbb{R}^N$ be a bounded open set. We define the Laplace operator with Dirichlet boundary conditions on $L_2(\Omega)$ by means of the continuous and coercive form $\mathfrak{a} : H_0^1(\Omega) \times H_0^1(\Omega) \to \mathbb{C}$ defined by

$$
\mathfrak{a}(u, v) = \int_\Omega \nabla u \overline{\nabla v} \, dx.
$$

For the relevant definitions and facts on forms we refer to [6, Section 5.3], [21, V.3] or [40, VI.§3]. The associated operator $-\Delta_2$ is invertible and sectorial of angle 0. Its domain is given by $\mathcal{D}(\Delta_2) = \{f \in H_0^1(\Omega) : \Delta f \in L_2(\Omega)\}$, where Δf is understood in the distributional sense. It is shown by Arendt and ter Elst [11, Section 3] that Δ_2 generates a holomorphic submarkovian C_0-semigroup (of self-adjoint operators) T_2 on $L_2(\Omega)$. Recall that a semigroup T on $L_2(\Omega)$ is called *submarkovian*, if each operator $T(t)$ is positive and $\|T(t)f\|_\infty \le \|f\|_\infty$ for all $f \in L_2(\Omega) \cap L_\infty(\Omega)$. Therefore, the semigroup extrapolates to holomorphic C_0-semigroups on $L_p(\Omega)$ for all $1 < p < \infty$; for the argument based on the Beurling-Deny criterion we refer to [6, Sections 7.1, 7.2], [11]. We denote the generator of T_p by Δ_p. The operators Δ_p inherit many properties of Δ_2, e.g. the spectrum $\sigma(\Delta_p)$ is independent of $p \in [1, \infty]$. In particular, all the operators Δ_p are invertible. The semigroup admits *Gaussian estimates*, that is the semigroup is given by a convolution kernel $K_t(x, y) \in L_\infty(\Omega \times \Omega)$ satisfying the estimate

$$|K_t(x, y)| \le ct^{-\frac{N}{2}} e^{b|x-y|^2 t^{-1}} e^{\omega t}$$

for all $t > 0$, where $b, c > 0$ and $\omega < 0$ are constants. In fact, the semigroup T_2 is dominated by the Gaussian semigroup [6, Ex. 7.4.1]. This has a number of interesting consequences.

In particular, for $1 < p < \infty$ the densely defined operators $-\Delta_p$ are sectorial of angle 0 and admit a bounded H^∞-functional calculus of the same angle (see [11, Thm. 5.7]). Consequently, the operators $-\Delta_p$ are R-sectorial of angle 0 and invertible. We deduce that Δ_p satisfies maximal regularity on the line (for the first order and second order Cauchy problem).

We will again consider multiplicative perturbations of the Dirichlet Laplacian Δ_p. For this reason we are interested in obtaining some information on the trace space Tr_q. As this space is strongly related with $(L_p(\Omega), \mathcal{D}(\Delta_p))_{\frac{1}{q^*}, q}$, we are interested to identify the domain $\mathcal{D}(\Delta_p)$. If Ω has a boundary of class C^2, it is shown in [21, IX.25] that $\mathcal{D}(\Delta_2) = H^2(\Omega) \cap H_0^1(\Omega)$; the same is known if the domain $\Omega \subset \mathbb{R}^N$ is convex and if $p = 2$ (see [1]). Unfortunately, for general non-smooth domains, this does not hold true, if $N > 1$. At least, we obtain as a consequence of the boundedness of the Riesz transform the embedding $\mathcal{D}(\Delta_p) \hookrightarrow W_0^{1,p}(\Omega)$ for $1 < p \le 2$ (see [15], [35]).

However, using the embedding properties of real interpolation spaces [115, p. 25] and the relation of domains of fractional powers of sectorial operators to real interpolation spaces between X and the domain space [115, p. 101 (3)], we have, for $\theta < \frac{1}{q^*} = 1 - \frac{1}{q}$, the following chain of continuous embeddings

$$(L_p(\Omega), \mathcal{D}(\Delta_p))_{\frac{1}{q^*}, q} \hookrightarrow (L_p(\Omega), \mathcal{D}(\Delta_p))_{\theta, 1} \hookrightarrow \mathcal{D}(A_p^\theta),$$

where $A_p = -\Delta_p$. Instead of the Sobolev embedding theorem will make use of the ultracontractivity of the semigroup in the following to deduce that $\mathcal{D}((-\Delta_p)^\theta) \hookrightarrow L_\infty(\Omega)$ for suitable $\theta > 0$. Indeed, the Gaussian estimates imply for $1 \le p < q \le \infty$ the estimate

$$\|T(t)\|_{p \to q} \le ct^{-\frac{N}{2}|\frac{1}{p} - \frac{1}{q}|} e^{\omega t}, \qquad t > 0,$$

for some constants $c > 0$ and $\omega < 0$ [6, Section 7.3]. Combining this estimate (for $q = \infty$) with the representation formula for negative fractional powers (see [61, Prop. 2.21]),

$$A_p^{-\theta} = \frac{1}{\Gamma(\theta)} \int_0^\infty t^{\theta-1} T_p(t) \, dt, \qquad \theta > 0,$$

we find that $A_p^{-\theta} : L_p(\Omega) \to L_\infty(\Omega)$ is bounded if $\theta > \frac{N}{2p}$ (integrability at zero). Since $\mathcal{D}(A^\theta) = \mathcal{R}(A^{-\theta})$, we have established

3.3.13 Lemma. *Let $\Omega \subset \mathbb{R}^N$ be an arbitrary open set, $p, q \in (1, \infty)$ and denote by Δ_p the Dirichlet Laplacian as defined above. Then we have the following continuous inclusion*

$$(L_p(\Omega), \mathcal{D}(\Delta_p))_{\frac{1}{q^*}, q} \hookrightarrow L_\infty(\Omega) \quad \text{if} \quad \frac{N}{2p} < 1 - \frac{1}{q}. \tag{3.24}$$

We consider the equation

$$\begin{cases} u_t \pm a(u)\Delta u = g, & (t, x) \in \mathbb{R} \times \Omega, \\ u(t, x) = 0, & (t, x) \in \mathbb{R} \times \partial\Omega. \end{cases} \tag{3.25}$$

Here, the function $a : \mathbb{C} \to \mathbb{R}$ is continuous, $a(0) > 0$, acts pointwise on u and g belongs to $L_q(\mathbb{R}, L_p(\Omega))$ with $1 < p, q < \infty$.

Now, we can mimic the proof of Theorem 3.3.11, using instead of the Sobolev embedding theorem the embedding (3.24). In this way we obtain the following related result.

3.3.14 Theorem. *Let $1 < p, q < \infty$ such that $\frac{N}{2p} < 1 - \frac{1}{q}$. Let $\Omega \subset \mathbb{R}^N$ be an arbitrary bounded open set and denote by Δ_p the Dirichlet Laplacian on Ω. Assume that the function a is Lipschitz continuous on bounded sets. Then, for $g \in L_q(\mathbb{R}, L_p(\Omega))$ sufficiently small, there is a solution u of $u_t \pm a(u)\Delta u = g$ satisfying*

$$u \in W^{1,q}(\mathbb{R}, L_p(\Omega)) \cap L_q(\mathbb{R}, \mathcal{D}(\Delta_p)).$$

Moreover, $u \in C_0(\mathbb{R}, \mathrm{Tr}_q) \hookrightarrow C_0(\mathbb{R}, L_\infty(\Omega))$.

The boundary condition is incorporated in the domain of the Dirichlet Laplacian.

Of course, an analogous result holds for the second order equation $u_{tt} + a(u)\Delta u = g$. In this case we can even improve the statement. Indeed, Lemma 3.3.6 allows to weaken the condition on the exponents p and q. We obtain the following theorem.

3.3.15 Theorem. *Let $1 < p, q < \infty$ such that $\frac{N}{2p} < 1 - \frac{1}{2q}$. Let $\Omega \subset \mathbb{R}^N$ be an arbitrary bounded open set and denote by Δ_p the Dirichlet Laplacian on Ω. Assume that the function a is Lipschitz continuous on bounded sets. Then, for $g \in L_q(\mathbb{R}, L_p(\Omega))$ sufficiently small, there is a solution u of $u_{tt} + a(u)\Delta u = g$, satisfying $(t, x) \in \mathbb{R} \times \Omega$*

$$u \in W^{2,q}(\mathbb{R}, L_p(\Omega)) \cap L_q(\mathbb{R}, \mathcal{D}(\Delta_p)) \cap C_0(\mathbb{R}, L_\infty(\Omega)).$$

We remark that, if $N = 3$ and $p = 2$, we are forced to choose $q > 4$ (first order) or $q > 2$ (second order).

A quasilinear operator in non-divergence form. We will consider the differential equations in non-divergence form

$$\frac{\partial^k}{\partial t^k}u + a(x, u(t,x))\Delta u = g, \qquad k \in \{1, 2\}, \tag{3.26}$$

on $\mathbb{R} \times \Omega$, where $\Omega \subset \mathbb{R}^N$ is an open bounded set, for Dirichlet or Neumann boundary conditions. We will assume that the function $a : \Omega \times \mathbb{C} \to \mathbb{R}$ is bounded, continuous and Lipschitz continuous in the second component (uniformly with respect to the first component) on all sets of the form $\Omega \times I$, where $I \subset \mathbb{C}$ is a bounded set, i.e. $|a(x, u) - a(x, v)| \leq L\,|u - v|$ for all $x \in \Omega$, and all $u, v \in I$. Moreover, assume that $m(x) = a(x, 0) > \alpha > 0$ for all $x \in \Omega$.

First we have to define a realization of the operator $m\Delta$ on $L_p(\Omega)$ and to show that it satisfies maximal regularity for the first and second order problem. Again, we proceed by means of forms. Consider the continuous, coercive and symmetric form $\mathfrak{a}(u, v) = \int_\Omega \nabla u \overline{\nabla v}$ with domain $V = H_0^1(\Omega)$ on the Hilbert space $L_2(\Omega, m^{-1}dx)$. The operator A_2 associated with the form a is $m\Delta_2$ and $\mathcal{D}(A_2) = \mathcal{D}(\Delta_2) = \mathcal{D}(m\Delta_2)$. Indeed, if $u \in \mathcal{D}(\Delta_2)$, $v \in V$, then we have

$$-\mathfrak{a}(u, v) = \int (\Delta_2 u)\,\overline{v} = \int (m\Delta_2 u)\,\overline{v}\,\frac{dx}{m}.$$

Because $m\Delta_2 u \in L_2(\Omega)$, we have $u \in \mathcal{D}(A_2)$ and $m\Delta_2 u = A_2 u$. The reverse inclusion follows analogously.

One verifies readily that the form $\mathfrak{a}$ satisfies the first and second Beurling-Deny criterion (see e.g. [6, Section 7.1]); in fact, the form does not depend on the function m. Therefore $m\Delta_2$ is the negative generator of a submarkovian positive analytic C_0-semigroup T_2 of self-adjoint operators on $L_2(\Omega, m^{-1}dx)$. Consequently, this semigroup extrapolates to positive analytic contractive C_0-semigroups T_p on $L_p(\Omega, m^{-1}dx)$ for $1 < p < \infty$. We denote the negative generators by A_p. We claim that $A_p = m\Delta_p$ with domain $\mathcal{D}(\Delta_p)$. Indeed, this is an immediate consequence of the identity $m\Delta_2 = A_2$ and the facts

1. $\mathcal{D}(A_2)$ is a core for A_p, i.e. $\mathcal{D}(A_p) = \overline{\mathcal{D}(A_2)}^{\mathcal{D}(A_p)}$ for $1 < p < 2$;

2. A_p is the part of A_2 in L_p, i.e. $A_p = A_2|_{L_p}$ for $2 < p < \infty$.

The first assertion holds, since the set $\mathcal{D}(A_2)$ is invariant under the semigroup T_2 and hence also under T_p; and because $\mathcal{D}(A_2)$ is dense in L_2 which embeds densely into L_p. The second assertion holds, as T_p is the restriction of T_2 to the invariant subspace L_p (see [54, p.52, p.60f] for details on the core and the part of an operator). Obviously, in case $2 < p < \infty$, the identity $A_p = m\Delta_p$ holds (recall that $\Delta_p = \Delta_2|_{L_p}$). In case $1 < p < 2$ observe that A_p and $m\Delta_p$ agree on the set $\mathcal{D}(\Delta_2)$ which is a core for A_p by 1. and also a core for $m\Delta_p$, since m induces an invertible operator.

In Example 1.5.11 we observed that the operator $-m\Delta_p$ has a bounded functional calculus on $L_p(\Omega, m^{-1}dx)$; its angle is less than $\pi/2$ (see [81, Cor. 10.16]). Since $m\Delta_p$ is invertible, we have maximal regularity on the line for the first and second order problem.

Remark that the norm induced by $m^{-1}dx$ is equivalent to the norm on $L_p(\Omega)$. We deduce that $m\Delta_p$ satisfies maximal regularity in $L_p(\Omega)$.

The desired maximal regularity result is also implied by a result of Lamberton and Weis [121]. Observe that we did not make use of any regularity of Ω up to now. However, regularity properties of the domain Ω may be required in order to obtain some information on the domain $\mathcal{D}(\Delta_p)$.

An alternative approach to this maximal regularity result can be based on the factorization $m\Delta = m^{1/2}(m^{1/2}\Delta m^{1/2})m^{-1/2}$, which shows that the operator $m\Delta$ is similar to an operator in divergence form, associated with the form $\mathfrak{b}(u,v) = \int \nabla(m^{1/2}u)\overline{\nabla(m^{1/2}v)}$ having domain $V_{\mathfrak{b}} = \{u : m^{1/2}u \in V\}$.

The identity $\mathcal{D}(m\Delta_p) = \mathcal{D}(\Delta_p)$ allows to make use of the fact that the semigroup $e^{t\Delta_p}$ is ultra-contractive. As in the preceding paragraph we obtain an embedding analogous to (3.24).

The operator family A in the abstract setting is given by $A(u) = a(\cdot, u)\Delta_p$. It follows from our assumptions on a that A is Lipschitz continuous on bounded sets $U \subset \mathrm{Tr}$. As $\mathcal{D}(b\Delta_p) = \mathcal{D}(\Delta_p)$ for bounded functions $b \in L_\infty(\Omega)$, we find that all the operators $A(u)$ have the common domain $D_A = \mathcal{D}(\Delta_p)$. Mimicking the arguments of the preceding paragraph we obtain the following theorem.

3.3.16 Theorem. *Let* $1 < p, q < \infty$, $k \in \{1, 2\}$ *be such that* $\frac{N}{2p} < 1 - \frac{1}{kq}$. *Let* $\Omega \subset \mathbb{R}^N$ *be an arbitrary bounded open set and denote by* Δ_p *the Dirichlet Laplacian on* Ω. *Let the function* a *be as above. Then, for* $g \in L_q(\mathbb{R}, L_p(\Omega))$ *sufficiently small, there is a solution* u *of* (3.26) *satisfying Dirichlet boundary conditions with*

$$u \in W^{k,q}(\mathbb{R}, L_p(\Omega)) \cap L_q(\mathbb{R}, \mathcal{D}(\Delta_p)) \cap C_0(\mathbb{R}, L_\infty(\Omega)).$$

In case $k = 1$ we may replace a by $-a$.

Choosing $V = H^1(\Omega)$ as the domain for the form $\mathfrak{a}$, we obtain a realization $m\Delta_p^N$ of the Laplacian with Neumann boundary conditions. The operator is in general not invertible, but it is so after a suitable translation. Hence, we have maximal regularity for the operator $m\Delta_p^N - \lambda$ for any $\lambda > 0$. If Ω has the *extension property* (in the sense of [6, Section 7.3.6], i.e. the restriction mapping $R : W^{1,2}(\mathbb{R}^N) \to W^{1,2}(\Omega)$ is surjective), we obtain again the inclusion (3.24). We summarize these considerations in the following theorem.

3.3.17 Theorem. *Let* $1 < p, q < \infty$, $k \in \{1, 2\}$ *be such that* $\frac{N}{2p} < 1 - \frac{1}{kq}$. *Let* $\Omega \subset \mathbb{R}^N$ *be an open set having the extension property and denote by* Δ_p *the Neumann Laplacian on* Ω. *Let the function* a *be as above. Then, for* $\lambda > 0$ *and* $g \in L_q(\mathbb{R}, L_p(\Omega))$ *sufficiently small, there is a solution*

$$u \in W^{k,q}(\mathbb{R}, L_p(\Omega)) \cap L_q(\mathbb{R}, \mathcal{D}(\Delta_p)) \cap C_0(\mathbb{R}, L_\infty(\Omega)).$$

of $\frac{\partial^k}{\partial t^k}u + a(x, u(t, x))\Delta u - \lambda u = g$, $k \in \{1, 2\}$, *satisfying Neumann boundary conditions.*

Elliptic operators. The Dirichlet Laplacian is only an instance of a class of more general second order elliptic differential operators (in divergence form) for which we can establish results of this kind. Indeed, we did only make use of the fact that the operator was invertible, R-sectorial of suitable angle and admitted Gaussian estimates. In [11] these properties were established for a whole class of second order elliptic operators with various boundary conditions. We will describe it in the following.

Let $\Omega \subset \mathbb{R}^N$ be a bounded open set. Let $a_{ij}, b_i, c_i, a_0 \in L_\infty(\Omega)$, $i, j \in \{1, \ldots, n\}$, be real-valued coefficients. We assume the ellipticity condition

$$\mathcal{R}e \sum_{i,j=1}^{N} a_{ij} \xi_i \bar{\xi}_j \geq \alpha \, |\xi|^2 \qquad \text{for all } \xi \in \mathbb{C}^n, \text{ a.e.}$$

where $\alpha > 0$. We will consider various realizations in L_2 of the elliptic differential operator in divergence form

$$Lu = -\sum_{i,j=1}^{N} D_i(a_{ij} D_j u) + \sum_{i=1}^{N} \big(b_i D_i u - D_i(c_i u)\big) + a_0 u.$$

This will be achieved again by means of forms. Let V be a closed ideal of $W^{1,2}(\Omega)$ containing $\mathring{W}^{1,2}(\Omega)$. We define the form $\mathfrak{a} = \mathfrak{a}_V : V \times V \to \mathbb{C}$ by

$$\mathfrak{a}(u,v) = \int_\Omega \Big[\sum_{i,j=1}^{N} a_{ij}(x) D_i u \overline{D_j v} + \sum_{i=1}^{N} \big(b_i D_i u \bar{v} + c_i u \overline{D_i v}\big) + a_0 u \bar{v} \Big] \, dx.$$

Then $\mathfrak{a}_V$ is a continuous, $L_2(\Omega)$-elliptic form and we associate to it the operator A_V. The operator $-A_V$ generates a holomorphic C_0-semigroup T_V on $L_2(\Omega)$.

The choice of $V = \mathring{W}^{1,2}(\Omega)$ corresponds to Dirichlet boundary conditions, the choice of $V = W^{1,2}(\Omega)$ to Neumann boundary conditions. One may also consider mixed boundary conditions (see [6, Section 8.1] for details).

Let V be either $\mathring{W}^{1,2}(\Omega)$ or $W^{1,2}(\Omega)$; in the second case assume that Ω has the extension property (see [6, 7.3.6]). It is shown in [11] that T_V extrapolates to holomorphic C_0-semigroups T_p in $L_p(\Omega)$ for $1 \leq p < \infty$ admitting Gaussian bounds. Denote the negative generator of T_p by A_p. Then, for $1 < p < \infty$, $A_p \in QH^\infty(\Sigma_\varphi)$ with $\varphi < \pi/2$ (see e.g. [11, Thm. 5.9], [6, Section 8.5] and [38]). We will assume in the following that

(H1) $i\mathbb{R} \subset \rho(A_p)$ or that

(H2) A_p is sectorial and invertible,

which can always be achieved by substituting a_0 by $a_0 + \omega$ for a suitable $\omega > 0$. We conclude that in this case A_p is R-sectorial of angle less than $\pi/2$ (given (H2)) or that A_p is R-bisectorial (given (H1)). Thus, A_p satisfies L_p-maximal regularity on the line for the first and second order Cauchy problem.

The Gaussian estimates imply again the embedding (3.24). Now we have all the required ingredients and we obtain by the same arguments as in the special case of the Dirichlet Laplacian the following result. We consider the equation

$$u_t \pm a(u)Lu = g, \qquad (t,x) \in \mathbb{R} \times \Omega, \tag{3.27}$$

where L is a realization A_p of the differential operator defined above satisfying assumption (H1), with Dirichlet or Neumann boundary conditions (in the second case we assume that Ω has the extension property). Moreover, the function $a : \mathbb{C} \to \mathbb{R}$ is continuous, $a(0) > 0$, acts pointwise on u and g belongs to $L_q(\mathbb{R}, L_p(\Omega))$ with $1 < p, q < \infty$.

3.3.18 Theorem. *Let $1 < p,q < \infty$ such that $\frac{N}{2p} < 1 - \frac{1}{q}$. Assume that the function a is Lipschitz continuous on bounded sets. Then, for $g \in L_q(\mathbb{R}, L_p(\Omega))$ sufficiently small, there is a solution u of* (3.27) *satisfying*

$$u \in W^{1,q}(\mathbb{R}, L_p(\Omega)) \cap L_q(\mathbb{R}, \mathcal{D}(A_p)).$$

Moreover, $u \in C_0(\mathbb{R}, L_\infty(\Omega))$.

This result extends to the second order case; we state it for reference. We assume that the realization A_p satisfies assumption (H2).

3.3.19 Theorem. *Let $1 < p,q < \infty$ such that $\frac{N}{2p} < 1 - \frac{1}{2q}$. Assume that the function a is Lipschitz continuous on bounded sets. Then, for $g \in L_q(\mathbb{R}, L_p(\Omega))$ sufficiently small, there is a solution of the equation $u_{tt} + a(u)Lu = g$ satisfying*

$$u \in W^{2,q}(\mathbb{R}, L_p(\Omega)) \cap L_q(\mathbb{R}, \mathcal{D}(A_p)) \cap C_0(\mathbb{R}, L_\infty(\Omega)).$$

A variational problem. Let I denote a bounded open interval in $\mathbb{R}$. We are interested in the following elliptic equation

$$u_{xx} + (a(u)u_y)_y = g, \qquad (x,y) \in \mathbb{R} \times I, \tag{3.28}$$

where $a : L_2(I) \to \{f \in L_\infty(I) : f \geq \alpha \text{ a.e.}\}$, $\alpha > 0$, is Lipschitz continuous on bounded subsets U of $L_2(I)$, i.e. $\|a(u) - a(v)\|_{L_\infty} \leq L \|u - v\|_{L_2}$ for all $u, v \in U$.

We will work with the Hilbert spaces

$$V = H_0^1(I) \hookrightarrow H = L_2(I) \hookrightarrow V' = H^{-1}(I),$$

where H^{-1} is the anti-dual space of H_0^1, i.e. the space of all continuous anti-linear forms on H_0^1. All the above embeddings are dense and continuous. In the notation of the abstract theorem, Theorem 3.3.8, put $X = H^{-1}(I)$ and $D_A = H_0^1(I)$. We define the maximal regularity space by

$$\mathrm{MR} = H^2(\mathbb{R}, H^{-1}(I)) \cap L_2(\mathbb{R}, H_0^1(I)),$$

then the trace space satisfies by Lemma 3.3.5

$$\mathrm{Tr} \subset (H^{-1}(I), H_0^1(I))_{1/2,2} \times H^{-1}(I) = L_2(I) \times H^{-1}(I)$$

with continuous inclusion. Given $w \in L_2(I)$, we consider the form $\mathfrak{b}_w$ on $H = L_2(I)$ with domain $D_A \times D_A$ given by

$$\mathfrak{b}_w(u, v) = \int_I a(w) u' \overline{v'} \, ds.$$

For the relevant definitions and facts on forms we refer to [6, Section 5.3], [21, V.3] or [40, VI.§3]. It is easy to see that the form $\mathfrak{b}_w$ is sesquilinear, symmetric, positive and continuous on L_2. Indeed, for $u, v \in D_A = H_0^1(I)$ we have

$$|\mathfrak{b}_w(u, v)| \leq \int_I |(a(w))(s) u'(s) v'(s)| \, ds \leq \|a(w)\|_{L_\infty} \|u\|_{H^1} \|v\|_{H^1} \, .$$

Now, [40, VI.§3 Thm. 6] states that the pairing

$$\mathfrak{b}_w(u, v) = \langle -A(w) u, v \rangle, \qquad \forall u, v \in V$$

defines a bounded operator $A(w) : V \to V'$, where the pairing is antiduality between V and V'. Moreover, the Poincaré inequality [21, VIII.12] tells us that the form $\mathfrak{b}_w$ is coercive. In fact, for all $u \in H_0^1(I)$ we have

$$\mathfrak{b}_w(u, u) \geq \alpha \|u'\|_{L_2}^2 \geq \alpha C \|u\|_{H_0^1}^2 \, .$$

Therefore, the theorem of Lax-Milgram [40, VI.§3 Thm. 7] implies that the operator $A(w) \in \mathcal{L}(V, V')$ is in fact an isomorphism. In particular we obtain the norm equivalence

$$\|A(w) \cdot\|_{V'} \sim \|\cdot\|_V \, .$$

Note that the domain V of the operator $A(w)$ does not depend on w. The operator $A(w) u$ corresponds to the expression $(a(w) u_y)_y$.

The operators $-A(w)$ are sectorial and invertible. Indeed, writing $A = A(w)$ shortly, observe first that for $\nu \geq 0$ the operator $-A + \nu$ is associated with the coercive form $\mathfrak{b}_w + \nu$; hence $[0, \infty) \subset \rho(A)$. It remains to show the resolvent estimate, i.e. we have to prove that there is a constant $M > 0$ such that

$$\left\|A(\nu - A)^{-1}\right\|_{V' \to V'} \leq M \qquad \forall \nu \geq 0.$$

Let $\phi \in V'$, then we have to show

$$\left\|A(\nu - A)^{-1} \phi\right\|_{V'} \leq M \|\phi\|_{V'} \qquad \forall \nu \geq 0.$$

Writing, $u = (\nu - A)^{-1} \phi$ this is equivalent to

$$M^{-1} \|u\|_V \leq \|(\nu - A) u\|_{V'} = \sup_{v \in V, \|v\|_V \leq 1} |\langle (\nu - A) u, v \rangle|$$

$$= \sup_{v \in V, \|v\|_V \leq 1} \left| \mathfrak{b}_w(u, v) + \nu(u, v)_{H_0^1} \right|,$$

for all $\nu \geq 0$. Now, the coercivity of the form $\mathfrak{b}_w + \nu$ implies for the choice $v = \|u\|_{H_0^1}^{-1} u$ (we may assume that $u \neq 0$)

$$c \|u\|_{H_0^1} \leq \|u\|_{H_0^1}^{-1} \left| \mathfrak{b}_w(u, u) + \nu(u, u)_{H_0^1} \right| \leq \|(\nu - A)u\|_{V'}$$

which proves sectoriality.

In particular, $A(0)$ satisfies maximal regularity for the problem $z'' + A(0)z = f$, since we are in a Hilbert space setting (see Theorem 3.2.1).

Let $U \subset L_2(I)$ be a bounded set. We will show that $A(\cdot) : U \to \mathcal{L}(H_0^1(I), H^{-1}(I))$ is Lipschitz continuous: let $u, v \in U \subset L_2(I)$ and $x \in H_0^1 = D_A$, then

$$\|(A(u) - A(v))x\|_{H^{-1}} = \sup_{y \in H_0^1, \|y\| \leq 1} |\langle (A(u) - A(v))x, y \rangle|$$

$$= \sup_{y \in H_0^1, \|y\| \leq 1} |\mathfrak{b}_u(x, y) - \mathfrak{b}_v(x, y)|$$

$$\leq \|a(u) - a(v)\|_{L_\infty} \|x\|_{H^1} \|y\|_{H^1} \leq L \|u - v\|_{L_2} \|x\|_{H^1}.$$

We summarize these considerations in the following theorem.

3.3.20 Theorem. *Let a be as above. Then for all $g \in L_2(\mathbb{R}, H^{-1}(I))$ small enough, there is a solution $u \in H^2(\mathbb{R}, H^{-1}(I)) \cap L_2(\mathbb{R}, H_0^1(I))$ of equation (3.28).*

Let us consider some examples of functions a satisfying the requirements of the above example. The easiest way to construct such a function is by taking the norm in an L_q space embedded in L_2, i.e. let a be given by $a(u) = 1 + \|u\|_{L_q} k$, where $1 \leq q \leq 2$ and $k \in L_\infty(I)$. Another example is $a(u) = \phi(\hat{u})$, where $\hat{u}$ denotes the Fourier transform of u (regarding $L_2(I)$ as a subspace of $L_1(\mathbb{R})$) and where $\phi : \mathbb{R} \to [1, \infty)$ is Lipschitz continuous on bounded intervals.

We will construct a last example. Let $I_1, \ldots, I_N$ be a partition of the interval I. Given a function $u \in L_2(I)$, denote by u_k the means $u_k = |I_k|^{-1} (\int_{I_k} |u|^2 \, dy)^{1/2}$. Denote by $\psi : L_2(I) \to L_\infty(I)$ the mapping $u \mapsto \sum_{i=1}^N u_k 1_{I_k}$. It is easy to verify that ψ is Lipschitz continuous; the Lipschitz constant depends only on the chosen partition of I. Then $a(u) = \phi \circ \psi(u)$, where $\phi : \mathbb{R} \to [1, \infty)$ is Lipschitz continuous on bounded intervals, defines a mapping from $L_2(I)$ into $L_\infty(I)$ that is Lipschitz continuous on bounded subsets of $L_2(I)$.

We may also include a term of the form $F(x, u) = h(x)u$, where $h \in L_2(\mathbb{R})$. The continuity of the embedding $\mathrm{Tr}_2 \hookrightarrow X \times X$ implies that F satisfies the required growth and Lipschitz estimates. Therefore, we obtain, for $\|g\|_{L_2}$ and $\|h\|_{L_2}$ sufficiently small, the existence of a solution $u \in H^2(\mathbb{R}, H^{-1}(I)) \cap L_2(\mathbb{R}, H_0^1(I))$ of the problem

$$u_{xx} + (a(u)u_y)_y + hu = g, \qquad (x, y) \in \mathbb{R} \times I.$$

3.4   Higher order equations

In this section we will generalize some of the results of the previous sections on first and second order equations.

Let X be a UMD-space and consider on $L_p(\mathbb{R}, X)$ the ordinary differential operator $A_k = \frac{d^k}{dx^k}$ with dense domain $\mathcal{D}(A_k) = W^{k,p}(\mathbb{R}, X)$ for $k \in \mathbb{N}$, $1 < p < \infty$. If we consider this operator as a Fourier multiplier, its symbol is given by $a_k(x) = (ix)^k I$ and the symbol of the resolvent $R(\lambda, A_k)$ is given by $(\lambda - (ix)^k)^{-1} I$.

We will apply the Mikhlin-Weis multiplier theorem to deduce that A_k are (bi-)sectorial operators. To this end we will use the following result on homogeneous functions.

3.4.1 Lemma. *Assume that $m \in C^1(\mathbb{R}^n \setminus \{0\})$ is (positively) homogeneous of degree $r \in \mathbb{R}$, i.e., $m(\rho\zeta) = \rho^r m(\zeta)$ for $\rho > 0$ and $\zeta \neq 0$, then each partial derivative of m is (positively) homogeneous of degree $r - 1$.*

We will obtain the desired sectoriality estimates by applying the Mikhlin-Weis theorem to the symbol of $\lambda R(\lambda, A_k)$.

3.4.2 Proposition. *Let X be a UMD-space and let $k \in \mathbb{N}$. Then*

1. *the operator $(-1)^k A_{2k}$ is sectorial of angle 0;*

2. *A_{2k-1} is bisectorial of angle 0, more precisely, $A_{2k-1} \in \mathrm{Sect}((-\pi/2, \pi/2), (0,0))$.*

Proof. The operators $(-1)^k A_{2k}$ are elliptic differential operators for which the result is well known (see [81, Thm. 6.2]). As A_1 generates the bounded translation group, we know that the statement holds in this case.

We will give a proof for the second affirmation. Assume that k is odd. The argument is similar to the one in the elliptic case. We need to establish the sectoriality estimate on the set $\Lambda = \{\lambda = e^{i\omega} |\sigma|^k : \omega \in J, \sigma \in \mathbb{R} \setminus \{0\}\}$, where $J \subset [0, 2\pi) \setminus \{\pi/2, 3\pi/2\}$ is a compact set (in the case k even we would have to exclude either 0 or π). Writing $\lambda = e^{i\omega} |\sigma|^k$, the symbol associated with $\lambda R(\lambda, A_k)$ is given by $m(\xi, \sigma, \omega) I$, where $m(\xi, \sigma, \omega) = e^{i\omega} |\sigma|^k (e^{i\omega} |\sigma|^k - a_k(\xi))^{-1}$, $(\xi, \sigma, \omega) \in (\mathbb{R}^2 \setminus \{0\}) \times J$. Clearly, $\xi \mapsto m(\xi, \sigma, \omega)$ is in $C^1(\mathbb{R} \setminus \{0\})$. We have to show that

$$\sup\{|m(\xi, \sigma, \omega)|, |\xi D_\xi m(\xi, \sigma, \omega)| : (\xi, \sigma, \omega) \in (\mathbb{R}^2 \setminus \{0\}) \times J\} < \infty.$$

As the map $(\xi, \sigma) \mapsto m(\xi, \sigma, \omega)$ is homogeneous of degree 0, (making use of the lemma above) this supremum is equal to

$$\sup\{|m(\xi, \sigma, \omega)|, |\xi D_\xi m(\xi, \sigma, \omega)| : (\xi, \sigma) \in \Gamma, \omega \in J\},$$

where $\Gamma = \{(\xi, \sigma) \in \mathbb{R}^2 \setminus \{0\} : \sigma^2 + \xi^2 = 1\}$ is compact. As the functions are continuous on the compact set $\Gamma \times J$, the supremum is finite. By Kahane's contraction principle Lemma 1.6.2 (for the R-bounds), we find that the symbol satisfies the Mikhlin condition uniformly in σ and ω, which proves the assertion. $\square$

The operators A_k have bounded H^∞-functional calculi with optimal angles.

3.4.3 Theorem. *Let X be a UMD-space and let $k \in \mathbb{N}$. Then*

1. *the operator* $(-1)^k A_{2k}$ *admits a bounded sectorial* $H^\infty(\Sigma_\theta)$-*functional calculus for all* $\theta > 0$.

2. A_{2k-1} *admits a bounded bisectorial functional calculus of angle* 0, *more precisely,* $A_{2k-1} \in H^\infty(\Sigma)$, *for all* $\theta > 0$, *where* $\Sigma = \Sigma_{(-\pi/2,\pi/2),(\theta,\theta)}$.

Proof. The cases $k = 1$ and k even (elliptic) are well known. The argument for odd k is similar. Let $f \in H_0^\infty(\Sigma)$, then the operator $f(A_k)$ is given by the contour integral

$$f(A_k) = \frac{1}{2\pi i} \int_\Gamma f(z) R(z, A_k) \, dz$$

where Γ is an admissible contour in $\Sigma \setminus i\mathbb{R}$. Taking the Fourier transform yields that $f(A_k)$ has the symbol $m_f I$, where

$$m_f(\xi) = \frac{1}{2\pi i} \int_\Gamma f(z)(z - a_k(\xi))^{-1} \, dz, \qquad \xi \in \mathbb{R}.$$

By Cauchy's theorem we have
$$m_f(\xi) = f(a_k(\xi)).$$

It is a consequence of Cauchy's theorem that $|tf'(it)| \leq c\|f\|_{H^\infty(\Sigma)}$ for all $t \in \mathbb{R}$, where the constant c does not depend on f. As $|\xi D_\xi m_f(\xi)| = |k\xi^k f'(i^k \xi^k)| \leq ck\|f\|_{H^\infty(\Sigma)}$, the symbol $m_f I$ satisfies the Mikhlin condition (by Kahane's contraction principle) and we obtain by the Mikhlin-Weis multiplier theorem that $\|f(A_k)\|_{\mathcal{L}(L_p(\mathbb{R},X))} \leq C\|f\|_{H^\infty(\Sigma)}$, i.e. A_k has a bounded bisectorial H^∞-functional calculus. $\qquad\square$

Analogously to the first and second order case we define the notion of maximal regularity of order k. Let $p \in (1, \infty)$ and $k \in \mathbb{N}$; consider the abstract Cauchy problem

$$u^{(k)}(t) + Cu(t) = f(t) \qquad (t \in \mathbb{R}), \tag{3.29}$$

where C is a densely defined closed linear operator on X, $u^{(k)}$ denotes the kth derivative with respect to $t \in \mathbb{R}$, and $f \in L_p(\mathbb{R}, X)$.

3.4.4 Definition. We say that C satisfies *maximal L_p-regularity* on the line for the Equation (3.29) if for all $f \in L_p(\mathbb{R}, X)$ there exists a unique solution $u_f \in W^{k,p}(\mathbb{R}, X) \cap L_p(\mathbb{R}, \mathcal{D}(C))$.

We may rephrase this by saying that C has L_p-maximal regularity if and only if the operator sum $A + B$ is bijective on $\mathcal{D}(A) \cap \mathcal{D}(B)$, where A is the derivation operator on $W^{k,p}(\mathbb{R}, X)$ and B is defined pointwise by $(Bu)(t) = C(u(t))$ on $L_p(\mathbb{R}, \mathcal{D}(C))$. It follows from the closed graph theorem that this is equivalent to the invertibility of $A + B$; in other words, the solution operator $f \mapsto u_f$ is an isomorphism from $L_p(\mathbb{R}, X)$ to $W^{k,p}(\mathbb{R}, X) \cap L_p(\mathbb{R}, \mathcal{D}(A))$.

An application of Theorem 1.7.7 yields immediately the following sufficient condition for maximal L_p-regularity.

3.4.5 Proposition. *Let X be a UMD-space, $1 < p < \infty$ and let $k \in \mathbb{N}$. The densely defined operator C satisfies maximal L_p-regularity for the k-th order Cauchy problem provided that either*

1. *k is odd and $i\mathbb{R} \subset \rho(C)$ with $\mathcal{R}\{tR(it, C) : t \in \mathbb{R}\} < \infty$;*

2. *$k = 2m$ is even and $I_m = (-1)^m(-\infty, 0] \subset \rho(C)$ with $\mathcal{R}\{tR(t, C) : t \in I_m\} < \infty$.*

3.4.6 Remarks.   1. Using this maximal regularity result, we may obtain corresponding results on quasilinear equations of higher order.

2. Note, that the situation is rather different, if one considers maximal regularity on an interval $[0, T)$, $T > 0$. In fact, in [57, Thm. 3.3] it is shown that, if the problem $u^{(k)} = Au$, where $k \geq 3$, is well-posed on $[0, \infty)$ (i.e. for all initial data there is a unique solution depending continuously on these data), then the operator A is necessarily bounded.

3.5   Examples of (multi)sectorial operators

In this short section we will illustrate that multisectorial operators arise naturally in applications. The prototype example for a sectorial operator is of course the Laplacian or, more generally, an elliptic differential operator. In the second chapter we encountered already the first derivative on $L_p(\mathbb{R})$ or, more generally, generators of bounded C_0-groups as examples of bisectorial operators. In Section 3.2.2 we obtained quite naturally bisectorial operators by reducing the second order Cauchy problem on the line to a first order system. Moreover, it is easy to construct bisectorial or even multisectorial operators by means of diagonal operators or L_p-multipliers (see Example 2.2.5 for an example in the bisectorial setting).

If a multisectorial operator has bounded spectral projections, we may decompose it into a finite number of sectorial operators. On the other hand, it is clear, that we may reverse this procedure and build multisectorial operators by taking direct sums of sectorial operators. For example, on the space $X = L_p(\mathbb{R})^N$ we obtain a multisectorial operator with domain $\mathcal{D}(A) = W^{2,p}(\mathbb{R})^N$ by defining $Af = A(f_1, \ldots, f_N) = (\alpha_1 \Delta f_1, \ldots, \alpha_N \Delta f_N)$, where $\alpha_k \in \mathbb{C} \setminus \{0\}$.

In the papers [55, 56] the authors study ordinary differential operators of the form

$$Ly = y^{(n)} + \sum_{k=1}^{n-1} a_k y^{(k)}$$

on $L_p(I)$, where I is a finite interval, $1 < p < \infty$ and $a_k \in L_p(I)$, equipped with so-called *regular* boundary conditions (see [56] for details). They show that L also gives rise to (asymptotically) multisectorial operators. The spectrum of L consists of at most countably many eigenvalues λ_k which cannot have a finite limit point, since L has compact resolvent. Furthermore, the numbers ρ_k, where $-\rho_k^n = \lambda_k$, lie with an error of $O(\frac{1}{\rho_k})$ (as

$|\rho_k| \to \infty$) on half-rays passing through the n-th roots of -1. Therefore, the eigenvalues λ_k lie within parabolas with axes $J_L = \{z : \arg z = \pm\frac{\pi n}{2}\}$, i.e. J_L is the positive real line for $n = 4q$, the negative real line for $n = 4q + 2$ and the imaginary axis for odd n (see [55, 56] for details). This situation is related to the one given by differential operators of higher order on the line which were considered previously in Section 3.4.

3.6 Existence of center manifolds

In this section we will illustrate another important application of maximal regularity to partial differential equations; we will establish a *reduction principle* that shows the existence of a finite dimensional center manifold containing all small bounded solutions. The methods employed in the proofs are similar to those employed studying quasilinear equations, i.e. we will formulate the problem as a fixed point problem. The smoothing property of maximal regularity will make Banach's fixed point theorem applicable.

In order to capture also bounded functions on the line we enlarge the Sobolev space, more precisely, we will work with Sobolev spaces endowed with an exponentially decaying weight. Given a Banach space X and $\alpha > 0$, we define the weighted Sobolev spaces

$$L_{p,\alpha}(X) = \{f : \mathbb{R} \to X : f \text{ measurable}, \|f\|_{p,\alpha} < \infty\},$$
$$W_\alpha^{1,p}(X) = \{f : \mathbb{R} \to X : f, f' \in L_{p,\alpha}(X)\},$$
$$\|f\|_{p,\alpha} = \left\|e^{-\alpha|\cdot|}f\right\|_p.$$

We define the norm in $W_\alpha^{1,p}(\mathbb{R}, X)$ by $\|f\|_{W_\alpha^{1,p}} = \|f\|_{\alpha,p} + \|f'\|_{\alpha,p}$. Observe that the relations $\|f(t) - f(s)\| \le |t - s|^{1-1/p} \|f'\|_{p,\alpha}$ and

$$f(0) = \int_{-1}^{0} [f(t) + (t + 1)f'(t)]\, dt \tag{3.30}$$

imply that the function f is continuous and that the mapping $f \in W_\alpha^{1,p}(X) \mapsto f(0) \in X$ is bounded.

We will consider an abstract quasilinear problem of the form

$$\frac{d}{dt}x - Lx = f(t, \lambda, x), \tag{3.31}$$

where L is a closed (unbounded) operator on a UMD Banach space X with dense domain $\mathcal{D}(L)$ and where $f : \mathbb{R} \times \Lambda \times \mathcal{D}(L) \to X$. Here $\Lambda \subset \mathbb{R}^n$ is a set of parameters.

We assume that the operator L and the underlying Banach space X decompose as described in the following. We have the decomposition $X = X_1 \oplus X_2$ where X_1 is finite dimensional; moreover, $L = A \oplus B$, $\mathcal{D}(L) = X_1 \oplus \mathcal{D}(B)$, where

(H1) $A \in \mathcal{L}(X_1)$ with $\sigma(A) \subset i\mathbb{R}$ and

(H2) B is invertible and R-bisectorial on X_2, i.e. $\mathcal{R}(\{sR(is, B) : s \in \mathbb{R}\})$ is finite.

Thus, we may rewrite equation (3.31) as

$$\frac{d}{dt}x_1 - Ax_1 = f_1(t, \lambda, x_1, x_2), \quad \frac{d}{dt}x_2 - Bx_2 = f_2(t, \lambda, x_1, x_2) \tag{3.32}$$

where $(x_1(t), x_2(t)) \in X_1 \oplus X_2$ and $f = (f_1, f_2)$. Denote by X_3 the space $\mathcal{D}(B)$ endowed with its graph norm; note that X_2 and X_3 are UMD spaces if and only if X is a UMD space. We will show that all (sufficiently small) bounded solutions of (3.32) lie on an integral manifold of the form

$$M_\lambda = \{(t, x_1, r(t, \lambda, x_1)) \in \mathbb{R} \times \mathcal{D}(L) : t \in \mathbb{R}, x_1 \in X_1\},$$

where $r : \mathbb{R} \times \Lambda \times X_1 \to X_3$ is continuous. A set $M_\lambda \subset \mathbb{R} \times X_1 \times X_3$ is called an *integral manifold* if, given $P = (s, \xi, r(s, \lambda, \xi)) \in M_\lambda$, there exits a solution u of $u' - Lu = f(\cdot, \lambda, u)$ with $u(s) = (\xi, r(s, \lambda, \xi))$ whose orbit is contained in M_λ, i.e. we have $(t, u(t)) \in M_\lambda$ for all $t \in \mathbb{R}$. (For more background on integral manifolds we refer to [26], [93] and the references therein).

The proof of the mentioned reduction theorem is based on a maximal regularity result for a system of first order equations. The following two lemmas are the key ingredients of this maximal regularity result. The first result takes care of the finite dimensional part.

3.6.1 Lemma. Let $\nu > 0$. For all $(\xi, g) \in X_1 \times L_{p,\nu}(\mathbb{R}, X_1)$ there is a unique solution $u \in W^{1,p}_\nu(\mathbb{R}, X_1)$ of the equation $u' - Au = g$, $u(0) = \xi$; moreover, the map $(\xi, g) \in X_1 \times L_{p,\nu}(\mathbb{R}, X_1) \mapsto u \in W^{1,p}_\nu(\mathbb{R}, X_1)$ is bounded.

Proof. It is clear, that the solution is unique; moreover it is given by the variation of constants formula:

$$u(t) = e^{tA}\xi + \int_0^t e^{(t-s)A}g(s)\, ds.$$

It is easy to see that the distributional derivative u' exists and that $u' = Au + g$; as A is bounded, it suffices to show that $u \in L_{p,\nu}$ with the estimate $\|u\|_{p,\nu} \le C(\|\xi\|_{X_1} + \|g\|_{p,\nu})$. The Jordan normal form of A gives at once that $\|e^{tA}\|$ is polynomially bounded, hence we may assume that $\xi = 0$. We estimate the integral using the triangle inequality

$$\|u\|_{p,\nu} \le I_1 + I_2 = \left(\int_{\mathbb{R}_+} \left\| e^{-\nu t} \int_{\mathbb{R}_+} e^{(t-s)A}g(s)1_{(0,t)}(s)\, ds \right\|^p dt \right)^{1/p}$$

$$+ \left(\int_{\mathbb{R}_-} \left\| e^{\nu t} \int_{\mathbb{R}_-} e^{(t-s)A}g(s)1_{(-t,0)}(s)\, ds \right\|^p dt \right)^{1/p}.$$

We will estimate the first term I_1 by the Young inequality for convolutions; the second term can be handled analogously.

$$I_1 = \left(\int_{\mathbb{R}_+} \left\| \int_{\mathbb{R}_+} e^{-\nu t}e^{(t-s)A}g(s)1_{(0,t)}(s)\, ds \right\|^p dt \right)^{1/p}$$

$$= \left(\int_{\mathbb{R}_+} \left\| \int_{\mathbb{R}_+} \left(e^{-\nu(t-s)}e^{(t-s)A}1_{(0,\infty)}(t-s) \right) \left(e^{-\nu s}g(s) \right) ds \right\|^p dt \right)^{1/p}$$

The function $h(s) = e^{-\nu s}g(s)1_{(0,\infty)}$ is an element of $L_p(\mathbb{R}, X_1)$. As all the eigenvalues of the matrix A are on the imaginary axis, the function $\left\|e^{tA}\right\|$ is polynomially bounded. Therefore, $k(s) = e^{-\nu s}e^{sA}1_{(0,\infty)}(s) \in L_1(\mathbb{R}, \mathcal{L}(X_1))$. Now, I_1 is just $\|k * h\|_{L_p}$, where $k * h$ denotes convolution. Hence, by Young's inequality (cf. [7, Prop. 1.3.5]) we have $I_1 \leq \|k\|_{L_1} \|h\|_{L_p}$. This proves the claim, since $\|h\|_{L_p} = \|g\|_{p,\nu}$. $\qquad\square$

The following lemma relates L_p-maximal regularity and maximal regularity in weighted spaces.

3.6.2 Lemma. *Let B be a closed operator on X_2 and assume that B satisfies maximal L_p-regularity on the line. Then there exists $\nu_0 > 0$ such that for all $\nu \in (0, \nu_0)$ and all $f \in L_{\nu,p}(\mathbb{R}, X_2)$ there is a unique $u_f \in W_\nu^{1,p}(\mathbb{R}, X_2)$ solving the equation*

$$u' - Bu = f. \tag{3.33}$$

Moreover, the application $f \in L_{\nu,p}(\mathbb{R}, X_2) \mapsto u_f \in W_\nu^{1,p}(\mathbb{R}, X_2)$ is bounded.

Proof. We denote by $S : L_p(\mathbb{R}, X_2) \to W^{1,p}(\mathbb{R}, X_2)$ the bounded solution operator $f \mapsto u_f$. This is where enters maximal L_p-regularity for B.

Let $\nu > 0$. Given $f \in L_{\nu,p}(\mathbb{R}, X_2)$, consider $\tilde{f} \in L_p(\mathbb{R}, X_2)$ defined by $\tilde{f}(t) = e^{-\nu|t|}f(t)$. The map $f \mapsto \tilde{f}$ is an isomorphism of the spaces $L_{\nu,p}(\mathbb{R}, X_2)$ and $L_p(\mathbb{R}, X_2)$. Now observe that u_f solves (3.33) in $L_{p,\nu}$ if and only if $\tilde{u}_f$ solves the equation

$$\tilde{u}' - B\tilde{u} = \tilde{f} - \nu(\mathrm{sign}\,(\cdot)\tilde{u}) \tag{3.34}$$

in classical L_p space. We apply the solution operator S (an isomorphism) on both sides and obtain

$$\tilde{u} = S(\tilde{u}' - B\tilde{u}) = S\tilde{f} - \nu S(\mathrm{sign}\,(\cdot)\tilde{u})$$

or equivalently

$$\tilde{u} + \nu T\tilde{u} = S\tilde{f},$$

where the operator T defined by $T\tilde{u} = S(\mathrm{sign}\,(\cdot)\tilde{u})$ is bounded with $\|T\| = \|S\|$ because $\|\mathrm{sign}\,(\cdot)\tilde{u}\|_{L_p} = \|\tilde{u}\|_{L_p} = \|u\|_{p,\nu}$. For $\nu < \|T\|^{-1}$ the operator $I + \nu T$ is invertible in $\mathcal{L}(L_p)$. Therefore, there exists a unique solution $\tilde{u}_f \in L_p$ of (3.34) and hence a unique solution $u_f \in L_{p,\nu}$ of (3.33). Moreover,

$$\|u_f\|_{p,\nu} + \|u_f'\|_{p,\nu} = \|\tilde{u}_f\|_p + \|\tilde{u}_f' + \nu\mathrm{sign}\,\tilde{u}_f\|_p$$
$$\leq (\nu+1)\|\tilde{u}_f\|_p + \|\tilde{u}_f'\|_p \leq C\left\|\tilde{f}\right\|_p = C\|f\|_{p,\nu},$$

which proves the boundedness of the solution operator. $\qquad\square$

Now we will combine the foregoing two lemmas to deduce from the assumptions (H1) and (H2) the following maximal regularity result.

3.6.3 Proposition. *Let L be a densely defined closed operator on the UMD space X. Assume that the assumptions (H1) and (H2) are satisfied. Then there exists $\nu > 0$ such that we have maximal regularity for the following system of first order equations.*

$$u_1' - Au_1 = g_1, \ u_1(0) = \xi, \tag{3.35}$$
$$u_2' - Bu_2 = g_2.$$

More precisely, for all $(\xi, g_1, g_2) \in X_1 \times L_{p,\nu}(X_1 \times X_2)$ there is a unique solution $(u_1, u_2) \in W_\nu^{1,p}(X_1) \times (W_\nu^{1,p}(X_2) \cap L_{p,\nu}(X_3))$ of the above system (3.35). Moreover, the solution operator $K : (\xi, g_1, g_2) \in X_1 \times L_{p,\nu}(X_1 \times X_2) \mapsto (u_1, u_2) \in W_\nu^{1,p}(X_1) \times (W_\nu^{1,p}(X_2) \cap L_{p,\nu}(X_3))$ is bounded.

Proof. The result about the second equation is an immediate consequence of the corresponding result on L_p-maximal regularity on the line (Theorem 3.1.5) and Lemma 3.6.2. The statement about the first equation is just Lemma 3.6.1. $\qquad\square$

The above proposition is the key-ingredient of the proof of the reduction theorem. This result was established by Mielke in the Hilbert space setting in [93] using maximal regularity results established in [92].

As the equation is autonomous, we can improve the above regularity result.

3.6.4 Lemma. *Let $(u_1, u_2) = K(\xi, g_1, g_2)$ with $(g_1, g_2) \in W_\nu^{1,p}(X_1 \times X_2)$. Then $(u_1, u_2) \in W_\nu^{1,p}(X_1 \times X_3)$ and*

$$\frac{d}{dt}(u_1, u_2) = K\left(A\xi + g_1(0), \frac{d}{dt}(g_1, g_2)\right).$$

Proof. Let $(v_1, v_2) = K\left(A\xi + g_1(0), \frac{d}{dt}(g_1, g_2)\right)$ and define $w_1(t) = \xi + \int_0^t v_1(s)\,ds$ and $w_2(t) = B^{-1}(v_2(0) - g_2(0)) + \int_0^t v_2(s)\,ds$. Then, $\frac{d}{dt}(w_1, w_2) = (v_1, v_2)$. Moreover, (w_1, w_2) is a solution of (3.35). Indeed,

$$w_1'(t) - Aw_1(t) = v_1(t) - A\xi - \int_0^t Av_1(s)\,ds = v_1(t) - A\xi - \int_0^t (v_1'(s) - g_1'(s))\,ds$$
$$= v_1(t) - A\xi - (v_1(t) - v_1(0) - g_1(t) + g_1(0)) = g_1(t),$$

since $v_1(0) = A\xi + g_1(0)$. Similarly, since $Bv_2 \in L_{p,\nu}$, we find

$$w_2'(t) - Bw_2(t) = v_2(t) - (v_2(0) - g_2(0)) + \int_0^t Bv_2(s)\,ds$$
$$= v_2(t) - v_2(0) + g_2(0) + \int_0^t (g_2'(s) - v_2'(s))\,ds = g_2(t).$$

Uniqueness of solutions forces $(w_1, w_2) = (u_1, u_2)$, which proves the stated formula and that $(u_1, u_2) \in W_\nu^{1,p}(X_1 \times X_3)$. $\qquad\square$

The argument below will make use of Lipschitz continuity of f; the smallness of the Lipschitz constant will provide strict contractivity as in the section on quasilinear equations. For this reason we impose the following condition on f.

(H3) $f = (f_1, f_2) \in C_{b,u}^1(\mathbb{R} \times \Lambda \times X_1 \times X_3, X_1 \times X_2)$ satisfies the following estimates

$$\|f(t, \lambda, x)\|_\infty = L_0, \quad \left\|\frac{\partial}{\partial t} f(t, \lambda, x)\right\|_\infty = L_1, \quad \left\|\frac{\partial}{\partial x} f(t, \lambda, x)\right\|_\infty = L_2,$$

where the suprema are taken over $\mathbb{R} \times \Lambda \times X_1 \times X_3$. The constants L_0, L_1, L_2 will be specified later; $C_{b,u}^1$ denotes the space of all (Fréchet) differentiable functions f such that f and its partial derivatives are bounded and uniformly continuous.

We state the reduction theorem.

3.6.5 Theorem. *Let X be a UMD space and assume that L and f satisfy the hypotheses (H1), (H2) and (H3). Let $p \in (1, \infty)$ and $\nu > 0$ as in Lemma 3.6.2.*
If L_0, L_1, L_2 are small enough, there is a function

$$r = r(t, \lambda, x_1) \in C(\mathbb{R} \times \Lambda \times X_1, X_3)$$

with the following properties:

1. *The set $M_\lambda = \{(t, x_1, r(t, \lambda, x_1)) \in \mathbb{R} \times \mathcal{D}(L) : t \in \mathbb{R}, x_1 \in X_1\}$, is an integral manifold for (3.32).*

2. *The orbit of every solution $u \in L_{p,\nu}(X_1 \times X_3)$ of (3.32) with $\lambda \in \Lambda$ is contained in M_λ, i.e. we have $(t, u(t)) \in M_\lambda$ for all $t \in \mathbb{R}$.*

This result was established in [93, Thm. 1] in the Hilbert space setting. The proof given there does not make use of the special structure of Hilbert space. The reason that Hilbert spaces were considered is that in our setting the maximal regularity result Proposition 3.6.3 was then unknown.

We need the following variant of the Banach fixed point theorem (see [4, p. 116]).

3.6.6 Lemma. *Let (X, d) be a complete metric space and Y be a topological space. Assume that $f : Y \times X \to X$ satisfies*

1. *There is a constant $\alpha \in (0, 1)$ such that $d(f(y, x_1), f(y, x_2)) \leq \alpha d(x_1, x_2)$ for all $x_1, x_2 \in X$ and $y \in Y$.*

2. *$f(\cdot, x) : Y \to X$ is continuous for all $x \in X$.*

Then, for each $y \in Y$, the map $f(y, \cdot) : X \to X$ has a unique fixed point $x(y)$ and $x(\cdot) \in C(Y, X)$.

In order to prove the smoothness of the solution obtained as a fixed point of a contraction we will apply the *fiber contraction theorem* that is formulated in the subsequent lemma.

Given a complete metric space E, we say that a mapping $T : E \to E$ has the *fixed point property*, if T has a unique fixed point and if, for each $x_0 \in E$, the sequence $x_n = T^n(x_0)$ converges to that fixed point.

3.6.7 Lemma. *Given two complete metric spaces E_1, E_2 and two mappings $T_1 : E_1 \to E_1$, $T_2 : E_1 \times E_2 \to E_2$, assume that T_1 has the fixed point property and that there is a constant $\alpha \in (0,1)$ such that $T_2(x, \cdot)$ is a contraction in E_2 with constant α for each fixed $x \in E_1$. Let $\bar{x}$ denote the fixed point of T_1 and $\bar{y}$ that of $T_2(\bar{x}, \cdot)$. Assume that $T_2(\cdot, \bar{y}) : E_1 \to E_2$ is continuous.*

Then the mapping $R : E_1 \times E_2 \to E_1 \times E_2$, $(x, y) \mapsto (T_1(x), T_2(x, y))$ possesses the fixed point property.

Proof. See [120, II.§13 IV]. Clearly, the fixed point of R is $(\bar{x}, \bar{y})$.   □

Proof of the reduction theorem. (Outline). Let MR $= L_{p,\nu}(X_1 \times X_3)$. We consider the mapping $S : \mathbb{R} \times \Lambda \times X_1 \times \text{MR} \to \text{MR}$ defined by

$$S(s, \lambda, \xi, u) = K\big(\xi, f(\cdot + s, \lambda, u(\cdot))\big)$$

where K is the solution operator of Proposition 3.6.3. Note that, as f is bounded, the function $f(\cdot + s, \lambda, u(\cdot))$ clearly lies in the weighted Sobolev space $L_{p,\nu}$.

A function $\bar{u} \in \text{MR}$ is a fixed point for $S(s, \lambda, \xi, \cdot)$ if and only if the function $v = \bar{u}(\cdot - s)$ solves equation (3.32) with $v_1(s) = \xi$; note that in this case $\bar{u}$ is in fact an element of $W^{1,p}_\nu(X_1) \times (W^{1,p}_\nu(X_2) \cap L_{p,\nu}(X_3))$.

The linear operator S is a strict contraction on MR. Indeed, let $u, v \in \text{MR}$, then, since $S(s, \lambda, \xi, u) - S(s, \lambda, \xi, v)$ is the unique solution of (3.35) for the right hand side $(\xi, g_1, g_2) = (0, f(\cdot + s, \lambda, u(\cdot)) - f(\cdot + s, \lambda, v(\cdot))$, we find

$$\|S(s, \lambda, \xi, u) - S(s, \lambda, \xi, v)\|_{\text{MR}} \leq M \|f(\cdot + s, \lambda, u(\cdot)) - f(\cdot + s, \lambda, v(\cdot))\|_{L_{p,\nu}(X_1 \times X_2)}$$

$$\leq M L_2 \|u - v\|_{L_{p,\nu}(X_1 \times X_3)}.$$

If L_2 is sufficiently small, then S is a strict contraction in the variable $u \in \text{MR}$, uniform and continuous with respect to the variable (s, λ, ξ). Therefore, by Lemma 3.6.6, $S(s, \lambda, \xi)$ has a unique fixed point $\bar{u}(s, \lambda, \xi)$ in MR that depends continuously on (s, λ, ξ).

The fixed point $\bar{u}(s, \lambda, \xi)$ belongs not only to $L_{p,\nu}(X_1 \times X_3)$ but even to $W^{1,p}_\nu(X_1 \times X_3)$ as we will show in the following. The idea is to formally differentiate $S(s, \lambda, \xi, u(\cdot))$ with respect to t and to replace the derivative du/dt by the variable z; this is motivated by Lemma 3.6.4. Define the mapping $T : \mathbb{R} \times \Lambda \times X_1 \times \text{MR} \times \text{MR} \to \text{MR}$ by

$$T(s, \lambda, \xi, u, z) = K\Big[A\xi + f_1\big(s, \lambda, \int_{-1}^0 [u(t) + (t+1)z(t)]dt\big),$$
$$\frac{\partial}{\partial t}f(\cdot + s, \lambda, u(\cdot)) + \frac{\partial}{\partial x}f(\cdot + s, \lambda, u)z(\cdot)\Big].$$

By (3.30) the integral expression in the argument of f_1 reduces to $u(0)$ if $u \in W^{1,p}_\nu(X)$ and if $z = \frac{du}{dt}$. If both u and $\frac{du}{dt}$ are in MR, we have by construction

$$\frac{d}{dt}S(s, \lambda, \xi, u) = T(s, \lambda, \xi, u, \frac{d}{dt}u). \tag{3.36}$$

Indeed, in this case $u \in W^{1,p}_\nu(X_1 \times X_3)$ and the identity follows from Lemma 3.6.4. If L_1 and L_2 are chosen sufficiently small, the mapping $T(s, \lambda, \xi, u, \cdot) : \text{MR} \to \text{MR}$ is again a strict contraction; denote the fixed point of $T(s, \lambda, \xi, \bar{u}(s, \lambda, \xi), \cdot)$ by $\bar{z}(s, \lambda, \xi)$. The mapping $T(\cdots, z) : \mathbb{R} \times \Lambda \times X_1 \times \mathrm{MR} \rightarrow \mathrm{MR}$ is continuous. Consequently, the function $\bar{z} : \mathbb{R} \times \Lambda \times X_1 \rightarrow \mathrm{MR}$ is continuous by Lemma 3.6.6; moreover, by Lemma 3.6.7 the iteration sequence $(u_{n+1}, z_{n+1}) = (S(\cdots, u_n), T(\cdots, u_n, z_n))$ converges to the fixed point $(\bar{u}, \bar{z})$ for every starting point $(u_0, z_0) \in \mathrm{MR} \times \mathrm{MR}$. Setting $(u_0, z_0) = (0, 0)$, we deduce from (3.36) that $z_n = \frac{d}{dt} u_n$ for all $n \in \mathbb{N}$, and hence $\bar{z}(s, \lambda, \xi) = \frac{d\bar{u}}{dt}(s, \lambda, \xi)$. Therefore, $\bar{u}$ is a continuous mapping from $\mathbb{R} \times \Lambda \times X_1$ into $W_\nu^{1,p}(X_1 \times X_3)$ and we may define $r(s, \lambda, \xi) = \bar{u}_2(s, \lambda, \xi)(0)$. The function r is continuous as the evaluation operator $u \in W_\nu^{1,p}(X_1 \times X_3) \mapsto u_2(0) \in X_3$ is bounded.

This yields the manifold $M_\lambda = \{(t, x_1, r(t, \lambda, x_1)) \in \mathbb{R} \times \mathcal{D}(L) : t \in \mathbb{R}, x_1 \in X_1\}$. It remains to verify that this manifold has the stated properties. This is done making use of the uniqueness of the fixed point $\bar{u}$. Indeed, $y(\cdot) = \bar{u}(s, \lambda, \xi)(\cdot - s)$ is a solution of (3.32) passing through $P = (s, \xi, r(s, \lambda, \xi))$ with $y'(\cdot + s) - Ly(\cdot + s) = f(\cdot + s, \lambda, y(\cdot + s))$; substituting the variable $\cdot$ by $\cdot + (t - s)$ we obtain that

$$y(\tau + t) = [K(y_1(t), f(\cdot + t, \lambda, y(\cdot + t)))](\tau)$$

for all $t, \tau \in \mathbb{R}$. The uniqueness of the fixed point $\bar{u}$ implies $y(\cdot + t) = \bar{u}(t, \lambda, y_1(t))(\cdot)$; evaluating at the point 0 we obtain

$$y_2(t) = \bar{u}_2(t, \lambda, y_1(t))(0) = r(t, \lambda, y_1(t)).$$

We conclude $(t, y_1(t), y_2(t)) \in M_\lambda$ for all $t \in \mathbb{R}$, i.e. M_λ is an integral manifold.

If u is a solution of (3.32) which lies in $L_{p,\nu}(X_1 \times X_3)$, then, by the uniqueness of $\bar{u}$, we have $u(t) = \bar{u}(t, \lambda, u_1(t))(0) = (u_1(t), r(t, \lambda, u_1(t)))$ and thus $(t, u(t)) \in M_\lambda$ for all $t \in \mathbb{R}$. □

It is possible to localize this construction by means of a suitable cut-off function. Assume that the function f satisfies the condition

(H3)' There are neighborhoods of zero $U_1' \subset X_1$, $U_3' \subset X_3$ such that

$$f = (f_1, f_2) \in C_{b,u}^1(\mathbb{R} \times \Lambda \times U_1' \times U_3', X_1 \times X_2).$$

Moreover, there is a value $\lambda_0 \in \Lambda$ such that

$$f(t, \lambda_0, 0) = \frac{\partial}{\partial x} f(t, \lambda_0, 0) = 0, \qquad t \in \mathbb{R}.$$

In the following we will construct a cut-off function $\varphi \in C^{1+\alpha}(X, [0, 1])$ that takes the value 1 on the unit ball B_X of the UMD space X and vanishes outside of $2B_X$. Let $\chi \in C^\infty(\mathbb{R}, [0, 1])$ with $\chi(t) = 1$ for $t \in [-1, 1]$ and $\chi(t) = 0$ for $|t| \geq 2$. Define the function $\varphi : X \rightarrow [0, 1]$ setting $\varphi(x) = \chi(\|x\|)$. The chain rule implies that the function φ is as smooth as the norm of X. Without loss of generality we may replace the norm by an equivalent norm. Therefore, it remains to establish the existence of an equivalent norm that is sufficiently smooth. Details on differentiability of norms and the subsequent results may be found in the comprehensive book [43]. If the Banach space X is reflexive, there is an equivalent C^1-norm (i.e. the norm is continuously Fréchet differentiable in $X \setminus \{0\}$). If X is a UMD space, then X is even super-reflexive and has hence an equivalent differentiable norm whose derivative is Hölder continuous of order α (for some $0 < \alpha \leq 1$) on the unit sphere of X (combine [43, IV.5.3] and [43, V.3.2]), consequently also on closed bounded sets that do not contain the origin.

Let X (and hence X_3) be a UMD space and consider, for $\epsilon > 0$ small enough, the function $f^\epsilon \in C^1_{b,u}(\mathbb{R} \times \Lambda \times X_1 \times X_3, X)$ defined by

$$f^\epsilon(t, \lambda, x) = f(t, \lambda, x) \chi(\epsilon^{-1} \|x_1\|_{X_1}) \chi(\epsilon^{-1} \|x_2\|_{X_3}) \chi(\epsilon^{-2} |\lambda - \lambda_0|);$$

renorm the spaces X_1 (Hilbert space) and X_3 such that the norms are in $C^{1+\alpha}$ for some $\alpha > 0$. Then, if ϵ is chosen sufficiently small, the function f^ϵ satisfies the assumption (H3) of the preceding theorem. The following theorem is proved by restricting the obtained manifold to a sufficiently small cylindrical domain. We only state it and refer to [93, Thm. 1] for a proof in the setting of Hilbert spaces.

3.6.8 Theorem. *Let X be a UMD space and assume that f and L satisfy the hypotheses (H1), (H2) and (H3)'. Then there are neighborhoods of zero $U_1 \subset U_1'$, $U_3 \subset U_3'$, a neighborhood $\Lambda_0 \subset \Lambda$ of λ_0 and a function $r \in C_b(\mathbb{R} \times \Lambda_0 \times U_1, U_3)$ such that the orbit of every solution of (3.32) with $\lambda \in \Lambda_0$ lying in $U_1 \times U_3$ belongs to the manifold $\{(t, x_1, r(t, \lambda, x_1)) : t \in \mathbb{R}, x_1 \in U_1\}$.*

In [93] one can find further properties of the reduction function r. It shares many properties with the non-linearity f; e.g. if f is periodic, then so is r. Moreover, if the Banach space has a $C^{k+1+\alpha}$-norm and if f is in $C^{k+1}_{b,u}$, then r is in C^k.

Examples of spaces with a C^∞-norm are Hilbert spaces or L_p spaces for even p. UMD spaces admit in general no C^∞-norm; for example, it is known that L_p has a C^2-norm if and only if $p \geq 2$ [43, V.1.1].

In [94] the reduction theorem is applied to *Saint Venant's problem*. This problem consists in finding elastic deformations of an infinite prismatic body taking given values for the cross-sectional resultants of force and moments (see [93, 94, 95, 96] and the references therein for details). The problem is first transformed into the abstract Hilbert space setting (3.31) and in [94, Thm. 2.2] the operator L is studied (in [94] it is denoted by $\mathscr{L}_{\mathscr{B}}$). It is shown that its spectrum consists only of discrete eigenvalues of vanishing real part, the space H_1 is the generalized kernel of L and of dimension six; moreover, by establishing a growth estimate for the resolvent $R(is, L), s \in \mathbb{R}$, one finds that the part $L|_{H_2}$ is a canonically bisectorial and invertible operator, which proves that the reduction theorem is applicable.

In the following paragraph we will consider two more examples of differential operators to which the reduction theorem is applicable.

Examples   1. Consider the non-linear elliptic equation

$$u_x + iu_y = f(x, \lambda, u), \quad (x, y) \in \mathbb{R} \times (0, 2\pi)$$

where u is periodic with period 2π in the second variable, i.e. $u(x, 0) = u(x, 2\pi)$ for all $x \in \mathbb{R}$. We may rewrite this equation as

$$u' + Au = f(t, \lambda, u)$$

where $A = i\frac{d}{dy}$ has domain $\mathcal{D}(A) = W^{1,p}_{\text{per}}(0, 2\pi)$ in $X = L_p(0, 2\pi)$, $1 < p < \infty$. It is well known that A has spectrum $\sigma(A) = \sigma_p(A) = \mathbb{Z}$ and that all eigenspaces are finite dimensional, as A has compact resolvent. The eigenspace corresponding to the eigenvalue k is one-dimensional and spanned by the eigenvector $f_k(y) = e^{-iky}$. Let $P \in \mathcal{L}(X)$ denote the spectral projection onto the space of constant functions X_1 and let $Q = I - P$. We may decompose the UMD-space $X = L_p(0, 2\pi)$ into the invariant spectral subspace corresponding to the eigenvalue zero and its complement, i.e. $L_p(0, 2\pi) = \mathbb{C} \oplus X_2$, where we identified $\mathbb{C}$ with the subspace of all constant functions in X, $X_2 = QX$. The operator $A|_{X_2}$ is densely defined, canonically bisectorial, invertible and has a bounded H^∞-functional calculus. In particular, it is canonically R-bisectorial.

If the nonlinearity is sufficiently smooth, for example the choice $f(x, \lambda, u) = u^2$ is possible, the reduction theorem is applicable and we obtain the existence of a reduction function $r : \mathbb{R} \times U \to QW^{1,p}_{\text{per}}(0, 2\pi)$, where $U \subset \mathbb{C}$ is a neighborhood of zero.

2. Consider the partial differential equation

$$u_{xx} + u_{yy} + u = f(x, \lambda, u), \quad (x, y) \in \mathbb{R} \times (0, \pi)$$

with boundary condition $u(x, y) = 0$ for $(x, y) \in \mathbb{R} \times \{0, \pi\}$. We rewrite this equation as a first order system. Let $\Delta = \Delta_y$ denote the Laplace operator with Dirichlet boundary conditions on $\Omega = (0, \pi)$ with domain $\mathcal{D}(\Delta) = W^{2,p}(\Omega) \cap \mathring{W}^{1,p}(\Omega)$ in $L_p(\Omega)$, $1 < p < \infty$. The Dirichlet Laplacian Δ has compact resolvent, its spectrum is given by $\sigma(\Delta) = \sigma_p(\Delta) = \{-k^2 : k \in \mathbb{N}\}$. $-\Delta$ is sectorial of angle 0 and has a bounded functional calculus of the same angle, hence it is R-sectorial of angle 0.

Let $A = -\Delta - 1$. Then A is sectorial and we may associate to it the operator matrix

$$\mathcal{A} = \begin{pmatrix} 0 & I \\ -\Delta - 1 & 0 \end{pmatrix}$$

acting on the space $X = \mathring{W}^{1,p}(\Omega) \times L_p(\Omega)$ that reduces the second order equation to a first order system. The operator matrix $\mathcal{A}$ has compact resolvent and its spectrum is just $\sigma(\mathcal{A}) = \sigma_p(\mathcal{A}) = \{\pm\sigma_k : k \in \mathbb{N}\}$, where $\sigma_k = (k^2 - 1)^{1/2}$. The corresponding eigenspaces are one-dimensional and spanned by $(\sin(k\cdot), \pm\sigma_k \sin(k\cdot))$, $k \in \mathbb{N}$. The spectral subspace X_1 corresponding to the imaginary axis is thus the eigenspace associated with the eigenvalue $\sigma_1 = 0$, i.e. $X_1 = \langle\sin\rangle \times \{0\}$. Let P denote the corresponding spectral projection and let $X_2 = (I - P)X$. We decompose the UMD space X into the direct sum of the invariant eigenspace X_1 its complement X_2. By Proposition 3.2.4 the operator $\mathcal{A}|_{X_2}$ is invertible and canonically R-bisectorial. If f is sufficiently smooth, the hypotheses of the reduction theorem are satisfied.

3.7   Notes and comments

Maximal regularity on the line has important applications in solving nonlinear equations, as we have seen in the previous sections. However, the theory of maximal regularity on the line is much less studied than its counterpart where one considers the abstract Cauchy problem on a finite interval or on the half-line (see for instance [25], [45], [81], [121] and the references contained therein).

The characterization of maximal regularity in terms of R-boundedness goes back to Weis [121]. We only know of [108, 109] and [104] where the authors employ the Mikhlin-Weis multiplier theorem in order to establish the sufficiency of R-bisectoriality for the first order problem on the line. The second order problem is even less considered than the first order problem, a reference being again [108, 109].

The approach to maximal regularity via the operator-sum method has been outlined already in [74]; the arguments given here adapt the ideas presented there. However, the reduction to the first order case and the occurring bisectorial operator-matrix seem to be new in this context. Also, the extension of the closed-sum theorem to the setting of asymptotically bisectorial operators and its application to the periodic problem seem to be new results.

Maximal regularity on a finite interval has been exploited in [27], [30] and [103] in order to obtain existence and uniqueness for local solution of quasilinear initial value problems. The applications to quasilinear equations on the line require a different setting. Yet, the approach that is used to tackle these problems is in the same spirit, i.e. the question is solved by transforming it into a fixed point problem.

Bibliography

[1] V. Adolfsson: L^p-integrability of the second order derivatives of Green potentials in convex domains. Pac. J. Math. **159**, 2 (1993), 201-225.

[2] D. Albrecht, X. Duong and A. McIntosh, Operator theory and harmonic analysis. In: *Instructional Workshop on Analysis and Geometry, Part III (Canberra, 1995)*, 77–136. Austral. Nat. Univ., Canberra, 1996.

[3] D. Albrecht, E. Franks and A. McIntosh, *Holomorphic functional calculi and sums of commuting operators*, Bull. Austral. Math. Soc. **58** (1998), no. 2, 291–305.

[4] H. Amann, *Gewöhnliche Differentialgleichungen*, de Gruyter, Berlin, 1983.

[5] W. Arendt, *Semigroups generated by elliptic operators*, Lecture notes for the TULKA Internet Seminar 1999/2000.

[6] W. Arendt, *Semigroups and Evolution Equations: Functional Calculus, Regularity and Kernel Estimates*, Handbook of Differential Equations, Evolutionary Equations, volume 1, Elsevier B.V., 2004.

[7] W. Arendt, C. J. K. Batty, M. Hieber and F. Neubrander, *Vector-valued Laplace transforms and Cauchy problems*, Birkhäuser Verlag, Basel, 2001.

[8] W. Arendt and S. Bu, *Tools for maximal regularity*, Math. Proc. Cambridge Philos. Soc., **134** (2003), 317–336.

[9] W. Arendt and S. Bu, *The operator-valued Marcinkiewicz multiplier theorem and maximal regularity*, Math. Z., **240** (2002), 311–343.

[10] W. Arendt and S. Bu, *Sums of bisectorial operators and applications*, Integral Equations and Operator Theory, to appear 2005.

[11] W. Arendt and A. F. M. ter Elst, *Gaussian estimates for second order elliptic operators with boundary conditions*, J. Operator Theory **38** (1997), no. 1, 87–130.

[12] W. Arendt, O. El-Mennaoui and M. Hieber, *Boundary values of holomorphic semigroups*, Proc. Amer. Math. Soc. **125** (1997), no. 3, 635–647.

[13] W. Arendt and B. de Pagter, *Spectrum and asymptotics of the Black-Scholes partial differential equation in (L^1, L^∞)-interpolation spaces*, Pacific J. Math. **202** (2002), no. 1, 1–36.

[14] P. Auscher, A. McIntosh and A. Nahmod, *Holomorphic functional calculi of operators, quadratic estimates and interpolation*, Indiana Univ. Math. J. **46** (1997), no. 2, 375–403.

[15] P. Auscher and P. Tchamitchian, *Square root problem for divergence operators and related topics*, Astérisque no. 249, 1998.

[16] W. G. Bade, *An operational calculus for operators with spectrum in a strip*, Pacific J. Math., **3**, (1953). 257–290.

[17] J. Bergh and J. Löfström, *Interpolation spaces. An introduction*, Springer-Verlag, Berlin, 1976, Grundlehren der Mathematischen Wissenschaften, No. 223.

[18] E. Berkson and T. A. Gillespie, *Spectral decompositions and harmonic analysis on UMD spaces*, Studia Math. **112** (1994), no. 1, 13–49.

[19] J. Bourgain, *Some remarks on Banach spaces in which martingale difference sequences are unconditional*, Ark. Mat. **21** (1983), no. 2, 163–168.

[20] J. Bourgain, *Vector-valued singular integrals and the H^1-BMO duality*, Probability theory and harmonic analysis (Cleveland, Ohio, 1983), Dekker, New York, 1986, pp. 1–19.

[21] H. Brézis, *Analyse fonctionnelle - Théorie et applications*, Masson, 1993.

[22] T. Burak, *On spectral projections of elliptic operators*, Ann. Scuola Norm. Sup. Pisa (3) **24** (1970), 209–230.

[23] T. Burak, *On semigroups generated by restrictions of elliptic operators to invariant subspaces*, Israel J. Math. **12** (1972), 79–93.

[24] D. L. Burkholder, *Martingale transforms and the geometry of Banach spaces*, Probability in Banach spaces, III (Medford, Mass., 1980), Springer-Verlag, Berlin, 1981, pp. 35–50.

[25] P. Cannarsa and V. Vespri, *On maximal L^p regularity for the abstract Cauchy problem*, Boll. Un. Mat. Ital. B (6) **5** (1986), no. 1, 165–175.

[26] C. Chicone, *Ordinary differential equations with applications*, Texts in Applied Mathematics, 34. Springer-Verlag, New York, 1999.

[27] R. Chill and S. Srivastava, *L^p maximal regularity for second order Cauchy problems*, preprint.

[28] P. Clément, B. de Pagter, F. A. Sukochev and H. Witvliet, *Schauder decomposition and multiplier theorems*, Studia Math. **138** (2000), no. 2, 135–163.

[29] P. Clément and S. Guerre-Delabrière, *On the regularity of abstract Cauchy problems and boundary value problems*, Atti Accad. Naz. Lincei Cl. Sci. Fis. Mat. Natur. Rend. Lincei (9), no. 4 (1998), 245–266.

[30] P. Clément and S. Li, *Abstract parabolic quasilinear equations and application to a groundwater flow problem*, Adv. Math. Sci. Appl. **3** (1993/94), no. Special Issue, 17–32.

[31] P. Clément and J. Prüss, *An operator-valued transference principle and maximal regularity on vector-valued L_p-spaces*, Evolution equations and their applications in physical and life sciences (Bad Herrenalb, 1998), Dekker, New York, 2001, pp. 67–87.

[32] P. Clément and J. Prüss, Some remarks on maximal regularity of parabolic problems. *Evolution equations: applications to physics, industry, life sciences and economics* (Levico Terme, 2000), 101–111.

[33] J. B. Conway, *Functions of One Complex Variable I*, Graduate Texts in Mathematics 11, Springer-Verlag, New-York, 1978.

[34] J. B. Conway, *A Course in Functional Analysis*, Graduate Texts in Mathematics 96, Springer-Verlag, New-York, 1990.

[35] T. Coulhon and X. T. Duong, *Riesz transforms for $1 \leq p \leq 2$*, Trans. Amer. Math. Soc. **351** (1999), no. 3, 1151–1169.

[36] M. G. Cowling, *Harmonic analysis on semigroups*, Ann. of Math. (2) **117** (1983), no. 2, 267–283.

[37] M. Cowling, I. Doust, A. McIntosh, and A. Yagi, *Banach space operators with a bounded H^∞ functional calculus*, J. Austral. Math. Soc. Ser. A **60** (1996), no. 1, 51–89.

[38] D. Daners, *Heat kernel estimates for operators with boundary conditions*, Math. Nachr. **217** (2000), 13-41.

[39] G. Da Prato and P. Grisvard, *Sommes d'opérateurs linéaires et équations différentielles opérationnelles*, J. Math. Pures Appl. (9) **54** (1975), no. 3, 305–387.

[40] R. Dautray and J. L. Lions, *Mathematical Analysis and Numerical Methods for Science and Technology*, Vol. 2, Springer-Verlag, Berlin (1988).

[41] E. B. Davies, *Heat kernels and spectral theory*, Cambridge University Press, Cambridge, 1990.

[42] R. Denk, M. Hieber and J. Prüss, *R-boundedness, Fourier multipliers and problems of elliptic and parabolic type*, Mem. Amer. Math. Soc., vol. 166 , Amer. Math. Soc, Providence, R.I., 2003.

[43] R. Deville, G. Godefroy and V. Zizler *Smoothness and renormings in Banach spaces*, Pitman Monographs and Surveys in Pure and Applied Mathematics, 64. Longman Scientific & Technical, Harlow; copublished in the United States with John Wiley & Sons, Inc., New York, 1993.

[44] J. Diestel, *Sequences and series in Banach spaces*. Springer-Verlag, New-York, 1984.

[45] G. Dore, *L^p regularity for abstract differential equations*, Functional analysis and related topics, 1991 (Kyoto), Springer-Verlag, Berlin, 1993, pp. 25–38.

[46] G. Dore, *H^∞ functional calculus in real interpolation spaces*, Studia Math. **137** (1999), 161–167.

[47] G. Dore, *H^∞ functional calculus in real interpolation spaces II*, Studia Math. **145** (2001), 75–83.

[48] G. Dore and A. Venni, *On the closedness of the sum of two closed operators*, Math. Z. **196** (1987), no. 2, 189–201.

[49] G. Dore and A. Venni, *Separation of two (possibly unbounded) components of the spectrum of a linear operator*, Integral Equations and Operator Theory, Vol. **12** (1989), 470–485.

[50] G. Dore and A. Venni, *H^∞ functional calculus for sectorial and bisectorial operators*, Studia Math. **166** (2005), 221-241.

[51] M. Duelli, *A characterization of Hilbert spaces by maximal regularity of Cauchy problems*, Diplomarbeit, Universität Ulm, 2000.

[52] M. Duelli, *Diagonal operators and L_p-maximal regularity*, Ulmer Seminare in Funktionalanalysis und Differentialgleichungen 2000, 156–173.

[53] M. Duelli and L. Weis, *Spectral projections, Riesz transforms and H^∞-calculus for bisectorial operators*, in preparation.

[54] K.-J. Engel and R. Nagel, *One-parameter semigroups for linear evolution equations*, Graduate Texts in Mathematics 194. Springer-Verlag, New York, 2000.

[55] I. D. Evzerov, *Domains of definition of fractional powers of ordinary differential operators in L_p spaces* (Russian), Mat. Zametki **21** (1977), no. 4, 509–518.

[56] I. D. Evzerov and P. E. Sobolevskiĭ, *Fractional powers of ordinary differential operators* (Russian), Differencial'nye Uravnenija **9** (1973), 228–240, 393.

[57] H. O. Fattorini, *Ordinary differential equations in linear topological spaces. I*, J. Differential Equations **5** (1969) 72–105.

[58] H. Föllmer, *Ein Nobel-Preis für Mathematik?*, Mitt. Dtsch. Math.-Ver. 1998, no. 1, 4–7.

[59] H. Grauert, K. Fritsche, *Einführung in die Funktionentheorie mehrerer Veränderlicher* Springer Hochschultext, Springer-Verlag, Berlin, 1974.

[60] P. Grisvard, An approach to the singular solutions of elliptic problems via the theory of differential equations in Banach spaces; in *Differential equations on Banach spaces*, Lecture Notes in Math. 1223, Springer-Verlag, Berlin, 131–155.

[61] M. Haase, *The functional calculus for sectorial operators and similarity methods*, doctoral dissertation, Universität Ulm, 2003.

[62] M. Haase, *Spectral properties of operator logarithms*, Math. Z. **245** (2003), no. 4, 761–779.

[63] M. Haase, *The Functional Calculus for Sectorial Operators*, Operator Theory: Advances and Applications, Birkhäuser, to appear.

[64] M. Haase, *A spectral mapping theorem for holomorphic functional calculi*, to appear in: J. London Math. Soc., 2005

[65] M. Haase, *A general framework for holomorphic functional calculi*, to appear in: Proc. Edin. Math. Soc., 2005.

[66] B. Haak, *Kontrolltheorie in Banachräumen mittels quadratischer Abschätzungen*, doctoral dissertation, Universität Karlsruhe, 2004.

[67] M. Hieber and J. Prüss, *Functional calculi for linear operators in vector-valued L^p-spaces via the transference principle*, Adv. Differential Equations **3** (1998), no. 6, 847–872.

[68] E. Hille and R. S. Phillips, *Functional analysis and Semigroups*. American Mathematical Society, Providence, Rhode Island, 1957.

[69] T. Hytönen, *Convolutions, multipliers, and maximal regularity on vector-valued Hardy spaces*, Helsinki University of Technology Institute of Mathematics Research Reports, preprint .

[70] J.-P. Kahane, *Some random series of functions*, Cambridge University Press, (1985).

[71] N. J. Kalton, *A remark on sectorial operators with an H^∞-calculus*, Trends in Banach spaces and Operator theory, 91-99 Contemp. Math. 321, AMS, Rhode Island (2003).

[72] N. J. Kalton, P. C. Kunstmann and L. Weis, *Perturbation and Interpolation Theorems for the H^∞-calculus with Applications to Differential Operators*, submitted.

[73] N. J. Kalton and G. Lancien, *A solution to the problem of L_p-maximal regularity* Math. Z. **235** (2000), no. 3, 559–568.

[74] N. J. Kalton and L. Weis, *The H^∞-calculus and sums of closed operators*, Math. Ann. **321** (2001), no. 2, 319–345.

[75] N. J. Kalton and L. Weis, *H^∞-functional calculus and square functions estimates*, in preparation.

[76] N. J. Kalton and L. Weis, *Euclidean structures and their applications to spectral theory*, in preparation.

[77] T. Kato, *Perturbation theory for linear operators*, Springer-Verlag, 1980.

[78] H. Komatsu, *Fractional powers of operators*, Pac. J. Math. **19** (1966), 285–346.

[79] P. C. Kunstmann and L. Weis, *Functional calculus and differential operators*. Lecture notes for the TULKA Internet Seminar 2001/2002.

[80] P. C. Kunstmann and L. Weis, *Perturbation theorems for maximal L_p-regularity*, Ann. Sc. Norm. Sup. Pisa **XXX** (2001), 415–435.

[81] P. C. Kunstmann and L. Weis, *Maximal L_p-regularity for Parabolic equations, Fourier multiplier theorems and H^∞-functional calculus*, in: Functional Analytic Methods for Evolution Equations, M. Ianelli, R. Nagel and S. Piazzera, eds., Lecture Notes in Mathematics, Springer, 2004.

[82] S. Kwapień, *Isomorphic characterization of inner product spaces by orthogonal series with vector-valued coefficients*, Studia Math. **44** (1972), 583–595.

[83] F. Lancien, G. Lancien and C. Le Merdy, *A joint functional calculus for sectorial operators with commuting resolvents*, Proc. London Math. Soc. (3) **77** (1998), no. 2, 387–414.

[84] G. Lancien, *Counterexamples concerning sectorial operators*, Arch. Math. (Basel) **71** (1998), no. 5, 388–398.

[85] C. Le Merdy, *H^∞-functional calculus and applications to maximal regularity*, In *Semi-groupes d'opérateurs et calcul fonctionnel*, Publ. Math. UFR Sci. Tech. Besançon. **16** (1998), 41–77.

[86] C. Le Merdy, *On square functions associated to sectorial operators*, Bull. Soc. Math. France **132** (2004), no. 1, 137–156.

[87] J. Lindenstrauss and L. Tzafriri, *Classical Banach spaces. I, Sequence spaces*, Springer-Verlag, Berlin, 1977, Ergebnisse der Mathematik und ihrer Grenzgebiete, Vol. 92.

[88] J. Lindenstrauss and L. Tzafriri, *Classical Banach spaces. II*, Springer-Verlag, Berlin, 1979, Function spaces.

[89] A. Lunardi, *Analytic semigroups and optimal regularity in parabolic problems*, Birkhäuser Verlag, Basel, 1995.

[90] A. McIntosh, *Operators which have an H_∞ functional calculus*, Miniconference on operator theory and partial differential equations (North Ryde, 1986), Austral. Nat. Univ., Canberra, 1986, pp. 210–231.

[91] A. McIntosh and A. Yagi, *Operators of type ω without a bounded H^∞-functional calculus*, Proc. Cent. Math. Anal. Aust. Natl. Univ. **24** (1990), 159–172.

[92] A. Mielke, *Über maximale L^p-Regularität für Differentialgleichungen in Banach- und Hilbert-Räumen*, Math. Ann. **277** (1987), no. 1, 121–133.

[93] A. Mielke, *Reduction of quasilinear elliptic equations in cylindrical domains with applications*, Math. Methods Appl. Sci. **10** (1988), no. 1, 51–66

[94] A. Mielke, *Saint-Venant's problem and semi-inverse solutions in nonlinear elasticity*, Arch. Rational Mech. Anal. **102** (1988), no. 3, 205–229.

[95] A. Mielke, *On Saint-Venant's problem for an elastic strip*, Proc. Roy. Soc. Edinburgh Sect. A **110** (1988), no. 1-2, 161–181.

[96] A. Mielke, *Corrigendum to: "Saint-Venant's problem and semi-inverse solutions in nonlinear elasticity"*, Arch. Rational Mech. Anal. **110** (1990), no. 4, 351–352.

[97] S. Monniaux, *Générateur analytique et régularite maximale*, Thèse, Université de Franche-Compté, Besançon, 1995.

[98] S. Monniaux, *A new approach to the Dore-Venni theorem*, Math. Nachr. **204** (1999), 163–183.

[99] A. Pazy, *Semigroups of linear operators and applications to partial differential equations*, Springer-Verlag, New York, 1983.

[100] G. Pisier, *Some results on Banach spaces without local unconditional structure*, Compositio Math. **37** (1978), no. 1, 3–19.

[101] G. Pisier, Les Inéqualités de Khinchine-Kahane, (d'après C. Borell) École Polyt. Palaiseau, *Sém. Géom. Espace de Banach* (1977/78), Exp.VII.

[102] J. Prüss, *Evolutionary Integral Equations and Applications*, Birkhäuser Verlag, Monographs in Mathematics, 87, 1993.

[103] J. Prüss, *Maximal regularity for evolution equations in L_p-spaces*, Conferenze del Seminario di Matematica dell'Universita di Bari, No. **285**, 1–39, Aracne Roma 2002.

[104] P. J. Rabier, *An isomorphism theorem for linear evolution problems on the line*, J. Dynam. Differential Equations **15** (2003), no. 4, 779–806.

[105] M. Riesz, *Sur les fonctions conjugées*, Math. Z. **27** (1928), 218–244.

[106] J. L. Rubio de Francia, *Martingale and integral transforms of Banach space valued functions*, Lecture Notes in Math. 1221 (1986), pp. 195–222.

[107] M. Reed and B. Simon, *Methods of modern mathematical physics II, Fourier analysis, Selfadjointness*, Academic Press (1975).

[108] S. Schweiker, *Asymptotics, regularity and well-posedness of first- and second-order differential equations on the line*, doctoral dissertation, Universität Ulm, 2000.

[109] S. Schweiker, *Mild solutions of second-order differential equations on the line*, Math. Proc. Cambridge Philos. Soc. **129** (2000), no. 1, 129–151.

[110] R. T. Seeley, *Complex powers of an elliptic operator*, Singular Integrals (Proc. Sympos. Pure Math., Chicago, Ill., 1966), Amer. Math. Soc., Providence, R.I., 1967, pp. 288–307.

[111] R. T. Seeley, *Norms and domains of the complex powers A_B^z*, Amer. J. Math. **93** (1971), 299–309.

[112] I. Singer, *Bases in Banach spaces I*, Springer-Verlag, Berlin, 1970.

[113] Ž. Štrkalj, *$\mathcal{R}$-Beschränktheit, Summensätze abgeschlossener Operatoren und operatorwertige Pseudodifferentialoperatoren*, doctoral dissertation, Universität Karlsruhe, 2000.

[114] Ž. Štrkalj and L. Weis, *On operator-valued Fourier multiplier theorems*, to appear in Trans. AMS

[115] H. Triebel, *Interpolation theory, function spaces, differential operators*, North-Holland Publishing Company 1978.

[116] H. Triebel, *Theory of function spaces*, Monographs in Mathematics, Vol. **78**, Birkhäuser 1983.

[117] M. F. Uiterdijk, *Functional calculi for closed linear operators*, doctoral dissertation, Universiteit Delft, 1998.

[118] A. Venni, *A counterexample concerning imaginary powers of linear operators*, Lecture Notes in Mathematics 1540, Springer-Verlag, Berlin (1993).

[119] A. Venni, *Banach spaces with the Hilbert transform property*, preprint.

[120] W. Walter, *Gewöhnliche Differentialgleichungen, Eine Einführung*, Heidelberger Taschenbücher, Band 110. Springer-Verlag, Berlin-New York, 1972.

[121] L. Weis, *A new approach to maximal L_p-regularity*, Evolution equations and their applications in physical and life sciences (Bad Herrenalb, 1998), Dekker, New York, 2001, pp. 195–214.

[122] L. Weis, *Operator-valued Fourier multiplier theorems and maximal L_p-regularity*, Math. Ann. **319** (2001), no. 4, 735–758.

[123] D. Werner, *Funktionalanalysis* 4. ed., Springer-Verlag, 2002.

[124] H. Witvliet, *Unconditional Schauder decompositions and multiplier theorems*, doctoral dissertation, Technische Universiteit Delft, November 2000.

[125] A. Yagi, *Coïncidence entre des espaces d'interpolation et des domaines de puissances fractionnaires d'opérateurs* C. R. Acad. Sci., Paris, Sér. I **299** (1984), 173–176.

[126] K. Yosida, *Functional Analysis*, Reprint of the sixth (1980) edition. Classics in Mathematics. Springer-Verlag, Berlin, 1995.

List of Symbols

Index

139

Zusammenfassung in deutscher Sprache

Diese Dissertation mit dem Titel "Functional calculus for bisectorial operators and applications to linear and non-linear evolution equations" befaßt sich mit dem Studium des Funktionalkalküls multisektorieller Operatoren und Anwendungen dieses Kalküls. Insbesondere betrachten wir maximale L_p-Regularität und nichtlineare Evolutionsgleichungen. Die Arbeit gliedert sich in drei Kapitel.

Gegenstand des ersten Kapitels ist die Entwicklung eines holomorphen Funktionalkalküls für multisektorielle Operatoren. In den ersten Abschnitten wird detailliert die Konstruktion beschrieben, und es werden viele Resultate, die für sektorielle oder bisektorielle Operatoren bereits bekannt sind, in dieser allgemeineren Situation bewiesen. In Sektion 1.7 verallgemeinern wir ein Resultat von Kalton-Weis. Dieser Satz liefert die Beschränktheit des operatorwertigen Funktionalkalküls für Funktionen mit R-beschränktem Bild. Daraus leiten wir dann sogenannte *Summensätze* her, die später für maximale Regularität wichtig werden. Wir erlauben dabei zum einen multisektorielle Operatoren und zum anderen *asymptotisch bisektorielle* Operatoren. Im letzten Abschnitt charakterisieren wir die Beschränktheit des Funktionalkalküls mittels *Quadratfunktionen* und des Begriffes der *fast R-Beschränktheit*. Dies hat einige interessante Konsequenzen: Zum Beispiel erhalten wir, daß ein bisektorieller[1] Operator A genau dann einen beschränkten bisektoriellen Funktionalkalkül hat, wenn die sektoriellen Operatoren $\pm iA$ einen beschränkten sektoriellen Funktionalkalkül besitzen. Dies gilt ohne jegliche Bedingung an die Geometrie des zugrundeliegenden Banachraumes.

Im zweiten Kapitel wenden wir die durch den Funktionalkalkül bereitgestellten Mittel an, um die Existenz und Eindeutigkeit von Spektralzerlegungen zu studieren. Es werden Charakterisierungen der Beschränktheit der Spektralprojektion durch fraktionäre Potenzen hergeleitet. Unter Benutzung der im vorangehenden Kapitel gezeigten Charakterisierung der Beschränktheit des Funktionalkalküls durch Quadratfunktionen konstruieren wir dann einen bisektoriellen Operator, dessen BIP-Typ größer oder gleich π ist. Ein Beispiel eines solchen Operators wurde bereits von Haase auf anderem Wege gefunden. Weiterhin leiten wir einige klassische Ungleichungen der Analysis aus der Beschränktheit der Spektralprojektion bzw. des Funktionalkalküls her. Im Fall, daß der bisektorielle Operator eine beschränkte C_0-Gruppe erzeugt, zeigen wir, daß die Existenz der Spektralzerlegung äquivalent zur Beschränktheit der verallgemeinerten Hilberttransformation ist. Dies verallgemeinert das klassische Beispiel der Translationsgruppe auf $L_p(\mathbb{R})$; der Generator dieser Gruppe ist die Ableitung. In den beiden letzten Abschnitten dieses Kapitels zeigen wir anhand einiger Beispiele, wie sich unter Ausnutzung der Beschränktheit der Spektralprojektionen Ergebnisse für sektorielle Operatoren auf multisektorielle Operatoren übertragen lassen.

---

[1] Ein Operator A heiße bisektoriell, wenn die beiden Operatoren $iA, -iA$ sektoriell sind.

Das letzte Kapitel ist dem Studium der maximalen L_p-Regularität und Anwendungen gewidmet. Zunächst charakterisieren wir die maximale L_p-Regularität auf $\mathbb{R}$ für das Cauchy-Problem erster und zweiter Ordnung. Dazu verwenden wir unter anderem die im ersten Kapitel gefundenen Summensätze. In dieser Situation treten in natürlicher Weise bisektorielle Operatoren auf. Wir zeigen auch, daß der Begriff des asymptotisch bisektoriellen Operators geeignet ist, um die maximale Regularität für das *periodische* Problem zu charakterisieren. In Abschnitt 3.3 beweisen wir zunächst abstrakte Resultate über die Existenz und Eindeutigkeit von Lösungen nichtlinearer Evolutionsgleichungen. Anschließend zeigen wir anhand einiger Beispiele, wie diese Ergebnisse auf konkrete partielle Differentialgleichungen anzuwenden sind. Der Beweis des abstrakten Resultats fußt auf der Beobachtung, daß maximale Regularität eine Glättungseigenschaft ist, die es erlaubt, Fixpunktargumente zu benutzen. Eine weitere Anwendung der maximalen Regularität geben wir in Sektion 3.6, wo wir einen auf Mielke zurückgehenden Satz über die Existenz von Zentrumsmannigfaltigkeiten von Hilberträumen auf UMD Banachräume übertragen.

Erklärung:

Hiermit erkläre ich, daß ich die Arbeit selbständig und nur mit den angegebenen Hilfsmitteln angefertigt habe. Alle Stellen, die anderen Werken entnommen sind, wurden durch Angabe der Quellen kenntlich gemacht.

Laupheim, den 06.01.2005